STRANGE MATTERS

ALSO BY TOM SIEGFRIED

The Bit and the Pendulum:
From Quantum Computing to M Theory—
The New Physics of Information

STRANGE MATTERS

undiscovered ideas at the frontiers of space and time

Tom Siegfried

Joseph Henry Press
Washington, D.C.

Joseph Henry Press • 2101 Constitution Avenue, N.W. • Washington, D.C. 20418

The Joseph Henry Press, an imprint of the National Academy Press, was created with the goal of making books on science, technology, and health more widely available to professionals and the public. Joseph Henry was one of the founders of the National Academy of Sciences and a leader in early American science.

Any opinions, findings, conclusions, or recommendations expressed in this volume are those of the author and do not necessarily reflect the views of the National Academy of Sciences or its affiliated institutions.

Library of Congress Cataloging-in-Publication Data

Siegfried, Tom, 1950-
  Strange matters : undiscovered ideas at the frontiers of space and time / by Tom Siegfried.
    p. cm.
Includes index.
  ISBN 0-309-08407-5 (cloth with jacket)
  1. Cosmology. 2. Physics. I. Title.
  QB981 .S535 2002
  523.1—dc21
                    2002006045

CONTENTS

PART TWO
STRANGE FRONTIERS

PART THREE
STRANGE IDEAS

PREFACE

I've always had a deep interest in the history of science. I find it fascinating and fun. And it seems to me that history should be an important part of telling any scientific story.

Some scientists agree. "I know of no better way of teaching science to undergraduates than through its history," the Nobel laureate physicist Steven Weinberg writes. "Science is, after all, part of the history of humanity."

If you replace "undergraduates" with "general public," Weinberg's sentiments are very similar to my own. And so in writing about science I like to draw on its history as much as I can.

But many books by many authors recount the stories from science's past. And I am not a historian, but a journalist. My job is usually to tell about science in the present.

In this book, though, I mix the present with the past and future. In a sense I'm trying to write the history of the science of the moment, reporting from the frontiers of research where history is in the process of being made. To do that in context I have to say something about the past. And then I try to go a step further, to tell about this history that hasn't yet been made.

I have focused on physics and cosmology because those fields provide two of the grandest mysteries for the future to solve. One is the nature of the "dark matter" that, astronomers have deduced, makes up most of the mass of the universe. The other is the question of how to reconcile Einstein's theory of gravity with quantum mechanics, the two most successful theories in the history of physics, yet seemingly incompatible with one another.

Put another way, the two mysteries might be expressed as two deep questions: "What is the universe made of?" and "How does the universe work?" In the course of trying to answer these questions, scientists have proposed the existence of many strange things: strange forms of matter, strange realms of existence, strange ideas about how time, space, and reality are related to one another. Most of these strange things strike the casual observer as bizarre beyond belief. And many, if not most, will no doubt turn out not to exist after all. But of all the strange matters I discuss in this book, I guarantee that at least one of them will someday be discovered. I just can't say which one.

My confidence comes from science's history, which is full of predictions of strange things that have come true. Mathematics possesses a mysterious power to reveal the existence of objects and phenomena before any physical evidence of their presence has been obtained. Just how that is possible is one of the main issues to be explored in the pages to follow.

Just how it was possible to write this book is another mystery that calls for some words of thanks. I extend deep appreciation to all of the scientists who have helped me over the years; most but not all will be mentioned in the text. And I thank my many friends and colleagues in the science journalism community who have politely listened to my verbalizations of this book before I began typing it. I hope they accept my apologies for not listing all of those who know their names deserve to be here.

My special thanks go to K.C. Cole of the *Los Angeles Times*, who read virtually the entire manuscript and ripped it to shreds with perceptive demands for clarity, logic, and completeness. Any science writer would be most unwise to publish a book without letting K.C. read it first.

K.C. is my second-favorite critic next to my wife, Chris, who's the best in the business at telling me when I don't make sense. (Close behind in this category are my science-journalist colleagues at the *Dallas Morning News*—Alexandra Witze, Sue Goetinck Ambrose, Laura Beil, and Karen Patterson.)

Thanks and sincere appreciation also go to my agent, Skip Barker, and my editor at Joseph Henry Press, Jeff Robbins, the key players in making this book real—even though it's a book about things that aren't real yet.

Tom Siegfried
Arlington, Texas
February 2002

INTRODUCTION

What is now proved was once only imagined.

—William Blake
  The Marriage of Heaven and Hell

Lisa Randall knows a thing or two about gravity.

On the one hand, as an avid rock climber, she regularly defies gravity's all-too-powerful pull, not always successfully. From first-hand experience she could testify that gravity can pose some serious problems.

On the other hand, as a physicist, she encounters a completely opposite problem with gravity. In the hierarchy of nature's forces, gravity is the weakling—magnetism is more than 100 times as strong, and the forces binding an atomic nucleus together are much stronger still. Gravity is the feeblest force in the universe, and for decades physicists have wondered why.

Randall knows no magic way to reduce the risks that gravity brings to rock climbing. But she does have a good idea for explaining why gravity on the cosmic scale is so weak.

1

For not only is Randall an intrepid explorer of national parks, she explores the universe as well. And since her spaceship is strictly mathematical, she doesn't have to worry about falling into a black hole or being blasted to smithereens by a supernova. And she doesn't have to restrict her travels to the space that we can see. Lisa Randall journeys to other dimensions. There she learns about the secrets of gravity and the existence of other worlds.

She is, in a way, a twenty-first-century version of A. Square, the protagonist of Edwin Abbott's nineteenth-century fantasy novel *Flatland*.

Abbott, a schoolteacher and theologian, described a totalitarian world whose inhabitants were like shadows on a sheet of paper, living very flat lives. Most men were polygons—triangles, squares, or hexagons, for example. High Priests were circles. Women were segments of straight lines.[1] The edges of all these figures were luminous so Flatlanders could see one another.

One day A. Square, a Flatlander mathematician, observed a small circle in his home. The circle would grow larger, expanding to a maximum width of about 13 inches, and then would contract again to a point before vanishing. This oscillating circle, A. Square eventually realized, was really a sphere from the third dimension, a realm of space previously unknown to the Flatlanders. Passing through the Flatlanders' sheet-of-paper universe, the sphere would become visible only where it intersected the sheet. And the intersection of a sphere with a plane is a circle.

When A. Square tried to explain the discovery of other dimensions to his fellow Flatlanders, they threw him in jail. A. Square's account of his ordeal was therefore dedicated with a plea to the outsideworlders who live in "space" to open their minds to the insights that extra dimensions had to offer.

Abbott's allegory has long been popular among mathematicians and scientists. But I doubt that the depth of the truth he revealed has been appreciated until very recently. For most of the past century, the common concept of an additional dimension stemmed from Einstein's theory of relativity, in which the "fourth dimension" is defined as time.[2] Time was not what Abbott had in mind, though. He clearly proposed the existence of higher dimensions of space, urging the citizens of "spaceland" to aspire to discovering the secrets of four, five, or even six spatial dimensions.

But surely, space itself has only three dimensions. Up and down, left and right, back and forth—three ways to move. Any movement can be described as some combination of movements in those three directions. Latitude, longitude, altitude—three dimensions—three numbers to specify the location of any object. What could be more obvious?

Only that it takes some imagination to see beyond the obvious. And that is what great scientists do.

A. Square saw beyond his two-dimensional world to a universe beyond. Lisa Randall sees beyond three-dimensional space to a realm of multiple dimensions, spaces that cannot be seen because light itself refuses to go there.

In fact, Lisa Randall's exploration of unknown dimensions shows how Abbott's prescient fantasy captures the essence of discovering the undiscovered—seeing beyond the horizons of the obvious. It may seem obvious to most people that space has three dimensions. But Abbott saw then, as Randall sees now, that exploring only the known dimensions restricts the prospects for great discoveries—and deeper understanding—about the universe and all that it contains. If all truths were obvious truths, there would be no need for science, or for scientists. But much of nature hides itself from human senses, rendering it unfamiliar when eventually revealed. Consequently, exploring the universe turns up some pretty strange things.

Scientists find those strange things in two places—out in space, and in their heads. Roughly speaking, those who seek novelty in space are called observers. Those who can work with their eyes closed are called theorists. There is an obvious symbiosis between these two species of scientist. When observers see something new in the cosmos, they call on theorists to explain it. When theorists dream up a new idea for something that the universe might contain, they expect observers to find it.

Theorists of the past have imagined many strange things in advance of their discovery by observers: antimatter, electromagnetic waves, black holes, and neutron stars; the expansion of the universe, the neutrino, quarks inside atoms, even atoms themselves. I like to call such instances of theoretical anticipation "prediscoveries." They suggest that science is richer and more creative than often presented—not mere observation, experiment, induction and deduction, but a process flush with creativity and imagination.

As have their intellectual ancestors, theorists of today have imagined many strange things that observers haven't yet found. Astrophysics journals, cosmology conferences, and World Wide Web pages are full of lengthy discourses on undiscovered objects and phenomena. Often the imaginations of scientists run completely unrestrained, and the resulting ideas bear no recognizable relation to standard science. Or even any thinkable future relation. But eliminating the flakes leaves plenty of exciting science on the edge—speculative yet plausible proposals about exotic objects that observers really do have a chance of finding someday.

Sometimes these proposals get a fair amount of media attention. For the most part, though, forecasts of new phenomena don't get the same respect as the odd things that actually have been found. But it's those ideas, those possible future discoveries, that transport us to the frontiers of the universe and point the way beyond. Strange ideas from theorists' imaginations guide observers in their efforts to learn more about matter, space, and time—and in fact, to explore the realm of reality beyond matter, space, and time.

To be sure, categorizing reality in terms of matter, space, and time has served science well. And in fact, even the boldest suggestions for novel phenomena usually build on the space-time-matter framework. Many historical examples of prediscovery—and some of the best candidates today—involve inferring the existence of new kinds of particles of matter, for example. It has become something of a physics tradition. Murray Gell-Mann imagined quarks years before evidence for them appeared in experiments. Decades earlier Wolfgang Pauli had invented a bizarre new particle called the neutrino, seemingly impossible to detect but essential for salvaging an important law of physics. Experimental proof of the neutrino's existence came a quarter century after Pauli's prediction.

Nowadays, inventing new forms of matter is a favorite pastime for physicists attempting to solve what astronomers call the "dark matter" problem. Based on the way galaxies spin and congregate in space, scientists can tell that the universe contains more matter than anybody can see. Perhaps it is ordinary matter, just not shining like stars. But few experts think so. Most of the evidence favors dark matter of some exotic flavor, probably made of particles of a species never detected on Earth. Inspired by the mystery of the dark matter—and by the elegance of certain equations—physicists have conceived of entire zoos of undiscovered particles that might very well pervade the cosmos.

Other dark matter candidates emerge from the work of physicists like Vic Teplitz, who seeks evidence of known matter particles in new disguises, perhaps in the form of "strange quark nuggets" that might be silently raining onto (and perhaps zipping through) Earth. While computer programmers tweak the software for his nugget-detecting "seismic telescope," Teplitz wanders through a looking glass to explore a "mirror world" of particles and stars. They are particles and stars that can't be seen but might be detected by the force of gravity—just the sorts of things that might make up the dark matter.

Further efforts to understand the cosmos provoke many of today's most magnificent prognostications. Great thinkers of the past, such as Alexander Friedmann, imagined a universe more dynamic than any scientist had before, a universe growing larger and larger. His vision was soon confirmed by observations. Today equally imaginative ideas suggest vast new frontiers of space and time for future scientists to explore. Most dramatically, scientists like Andrei Linde and Alan Guth have conceived of countless new universes bursting into existence far beyond our view. Guth and Linde are among many advocates of the view that our universe isn't the only one—though perhaps it's the best one, for living things like us.

But a multiplex of universes isn't the end of the story. More new ideas are needed to explain the universe we already know about. Another cosmotheorist, Paul Steinhardt, is a leader in the effort to explain why our universe seems to be expanding faster now than it used to be. He and others propose new versions of a ubiquitous cosmic fluid that may occupy every groove and wrinkle of all of space.

Besides identifying dark matter and understanding the cosmos, a third grand motivation inspires many attempts at prediscovery: the urge to unify science's theories of matter and force. Such efforts celebrate a noble tradition, exemplified by James Clerk Maxwell's nineteenth-century theory of electromagnetism. By figuring out the math that simultaneously described both magnetism and electricity, Maxwell prediscovered radio waves—as well as other forms of electromagnetic rays that transformed human lifestyles in the twentieth century. Today every physicist hopes to see the day when electromagnetism and the forces of the atomic nucleus are mathematically joined with gravity. The road to that unification leads toward a foggy horizon, but the pioneers suspect that their journey will reveal ultratiny objects called superstrings—or perhaps something even stranger.

Already, it seems, that road to unification twists and turns in some of those new dimensions of space explored by Lisa Randall and others, like Stanford's Savas Dimopoulos. (Always an enthusiast for new ideas, Dimopoulos promotes those extra dimensions as a place where Bill Gates might someday want to store information.) And space's secrets may get even stranger. Other authorities proclaim that the universe may be wrapped around itself in such a way as to create ghost images of every galaxy. If so, a distant patch of light on the night sky might just be the backside of our own galaxy, the Milky Way. And if that's not bizarre enough, free thinkers like Harvard's Cumrun Vafa would like to tell you that space is not the only place where dimensions should be added; the world may be big enough for more than one dimension of time as well.

It should go without saying that not all of the visions at the frontiers of physics will turn out to be true. Some are mirages, propelling the pioneers forward toward disappointment. But forward nonetheless. And surely some of the airy visions will soon solidify, at least if the successes of the past are any guide.

Therein lies the central mystery, however. Most of the pre-discoveries of the past have exploited the power of mathematics to represent reality, even parts of reality that have until then never been seen. How can math do that? Or as my friend Rosie Mestel says, how is it that squiggles on paper can tell us of the existence of things in the real world never before encountered?[3]

Usually this question is posed in terms of "the unreasonable effectiveness of mathematics," from a famous 1960 paper by the physicist Eugene Wigner. He emphasized the ability of scientists to take the crude observations of complicated experience and extract equations capturing regularities within the complexity. Perhaps, he suggested, such success is achieved because the typical physicist is rather irresponsible.

"When he finds a connection between two quantities which resembles a connection well-known from mathematics, he will jump at the conclusion that the connection *is* that discussed in mathematics," Wigner comments. "The mathematical formulation of the physicist's often crude experience leads in an uncanny number of cases to an amazingly accurate description of a large class of phenomena."[4]

But Wigner was talking about the success of theories after the fact, so to speak. He marvels at the precision of quantum mechanics for computing the energy of electrons in a helium atom, even though Werner Heisenberg's original quantum equations were based on properties that helium does not possess. Somehow, Wigner says, we "got something out of the equations that we did not put in."[5] But he does not discuss the power of mathematics to reveal previously unimagined phenomena. Apparently he found the ability of math to work at all mysterious enough.

In a recent book, though, two cognitive scientists—a psychologist and a linguist—argue that Wigner's mystery is illusory. Mathematics, these authors say, is simply an invention of the human mind, based on common human experience. There is no idealized, "platonic" mathematics inherent in nature; math is what people make it.

"It is sometimes assumed that the effectiveness of mathematics as a scientific tool shows that mathematics itself exists *in the structure of the physical universe*," these scientists write.[6] "This, of course, is not a scientific argument with any empirical scientific basis."

Oh? I would say the "of course" in the previous sentence lacks substantial justification. But these scientists argue vigorously in their book that math has nothing to do with the way the universe works apart from human descriptions of it.

"All the 'fitting' between mathematics and the regularities of the physical world is done *within the minds of physicists who comprehend both*," the cognitive scientists assert. "The mathematics is in the mind of the mathematically trained observer, not in the regularities of the physical universe."[7]

So there is no mystery, they say. We impose our math on the world in order to describe it. That's why math works.

Frankly, I am not impressed by this argument. Although it is surely true, at least in some senses, that math is a human invention, it does not logically follow that the universe does not live by mathematical laws. The idea of math as merely a human invention may explain much of its success. But I do not see how it explains the way that math reveals unseen, even unimagined, features of reality. It's one thing to fit equations to aspects of reality that are already known; it's something else for that math to tell of phenomena never previously suspected. When Paul Dirac's equations describing electrons produced more than one solution, he surmised that nature must possess other particles, now known as antimatter. But scientists did not discover such particles until after Dirac's math disclosed their existence. If math is a human invention, nature seems to have already known what was going to be invented.

It may well be true that humans have built mathematics out of concepts drawn from human experience. Yet somehow that realization does not resolve the mystery about why math works so well, but rather deepens it. For even with all the latest advances in brain science—revealing how humans think and reason and understand their environment—math's power to predict the reality of strange matters remains unexplained. But perhaps exploring the prediscoveries of the past and the potential prediscoveries of today can provide some clues to that mystery.

PART ONE

STRANGE MATTERS

STRANGE MATTER

1

From Gell-Mann and Quarks to the Search for Quark Nuggets

*The continued existence of the Moon, in the form we know it, despite
billions of years of cosmic-ray exposure, provides powerful empirical
evidence against the possibility of dangerous strangelet production.*

—R. Jaffe, W. Busza, F. Wilczek, and J. Sandweiss
Reviews of Modern Physics

As the second millennium of the Christian era ended, many people
feared that the world would, too.

Some anticipated Armageddon. Others were obsessed with Y2K.
And a few of the more scientifically minded among the worriers
dreaded the impending creation of strange quark matter.

Of course, strange quark matter had nothing to do with the end
of the millennium. It was mere coincidence that a powerful new atom
smasher on Long Island was scheduled to be up and running by late
1999. (And the millennium didn't really end until the end of the year

2000, anyway.) Nevertheless, in the months before its inaugural collisions, whispers began to spread that the new machine possessed the power to destroy the Earth—if not the whole universe.

Some of the whispers made it into print in popular media, as in a *Newsweek* article about the atom smasher, the Relativistic Heavy Ion Collider at the Brookhaven National Laboratory. Concerned with public reaction to such disaster rumors, Brookhaven's director, John Marburger, appointed four premier physicists to analyze any threat that the collider, known as RHIC (pronounced "Rick"), might actually pose to the planet.

By October 1999 the experts posted their report on the Internet for anyone to see. Their analysis identified three possible catastrophes: RHIC might create a small black hole (that would suck the Earth in); RHIC could "freeze space" throughout the universe, wiping out everything that had existed up till then; or RHIC might accidentally create a "strangelet," a small lump of matter made of an unusual mix of quarks.

By far, the strangelet scenario was the most serious to fear. The other two derived from an utter misconception about RHIC's real power.

RHIC acquired its risky reputation because it would be smashing together atoms of gold—hundreds of times heavier than the particles commonly smashed in such machines. Therefore the energy generated in the collisions would be higher than in any previous subatomic experiments. Fear of that unknown energy realm provoked predictions of apocalypse. With more energy available than ever before, perhaps RHIC *would* produce a black hole capable of swallowing the Earth. Or worse, perhaps the vacuum of space is not in the most stable possible state but is on the edge of transformation, like a supercooled liquid, poised to freeze at the slightest disturbance. Maybe RHIC's collisions would disrupt the fragile condition of space itself, sparking a phase transition that would in essence cause the whole universe to go poof!

But Brookhaven's expert panel dismissed those concerns. What's important is not the total energy, but the concentration of energy. Lower-energy collisions with smaller particles have already produced higher-energy concentrations than RHIC will with the relatively big gold atoms. Total energy is simply not the right thing to worry about. "If it were," the RHIC panel wrote, "a batter striking a Major League fastball would be performing a far more dangerous experiment than any contemplated at a high-energy accelerator."[1]

Therefore the black hole concern can be readily dismissed. Generating a black hole requires an enormous concentration of energy in a very small region of space. RHIC's energy would fall short by at least a factor of 10 billion trillion. As for triggering a sudden "freezing" of the vacuum of space, cosmic rays have already collided in deep space many times with a greater energy concentration than RHIC could muster. So the vacuum of space is surely stable enough to withstand any blips on RHIC's energy scale.

On the other hand, the strange quark matter scenario was a little more disturbing, perhaps because RHIC's very purpose was to create a new state of matter, best described as "quark soup."

THE UNIVERSE AS SOUP KITCHEN

Unless you were around to witness the birth of the universe, you've never tasted quark soup. Cooking it up requires temperatures something like a billion times hotter than in the sun, higher than anything the universe has seen since a few millionths of a second after the big bang. To generate that much heat, RHIC flings gold atoms through a ring 2.4 miles around, smashing the atoms together at nearly the speed of light, cramming the matter tightly enough to reach densities 30 times greater than a gold atom's nucleus (comparable to cramming all the matter in the moon into a ball that would fit in a backyard swimming pool). The dense heat melts the gold nuclei, squeezing out the quarks and gluons inside to create the soup of a time long past.

Technically minded physicists call quark soup a "quark-gluon plasma"; it is, after all, more like a gas than a liquid and does contain gluons as well as quarks. In today's cooled-down universe, quarks are the main building blocks of matter, congregating in triplets to form the protons and neutrons of an atom's nucleus. Gluons are the nuclear equivalent of Velcro, forcing the quarks within each nuclear particle to stick together.

But in the good old (old, old) days, when the universe was hot enough, quarks and gluons flowed freely through the plasma-soup. Only when the universal thermostat dropped below 10 trillion degrees did the quarks coagulate into protons and neutrons.

Along with the electrons (which are quark-free), protons and neutrons make up all the ordinary matter of everyday life, such as rocks and people, water and popcorn. Until mid-way through the twentieth century, protons and neutrons were regarded as the "uncuttable" components of the atomic nucleus, more similar to the ancient Greek concept of atoms than atoms themselves. But then along came a man who used math to see inside protons and neutrons—a guy who drives around today in an SUV with a license plate that reads QUARKS.

His name is Murray Gell-Mann, and he became the driving force in subatomic physics in the 1950s, a time when physicists were bewildered by all the particles they were discovering.

PHYSICS VERSUS BOTANY

Leon Lederman, a Nobel winner in the particle physics game, recalls those days as a time of particle plenty. "It was almost routine: You set up your apparatus in front of a new machine and you found a particle and you passed Go and you collected two hundred dollars," Lederman reminisced when I visited him in 1997 at the Fermi National Accelerator Laboratory in Batavia, Illinois (popularly known as Fermilab, home of the nation's most powerful atom smasher). In those early days, particles proliferated too rapidly even for Enrico Fermi, one of the century's greatest physicists.

"We were standing in lunch line, and I had to make conversation with the great man," Lederman recalled from a conference in the fifties. "I said, 'What do you think of the evidence for the V-zero-2 particle' that was just presented, and he looked at me and said, 'My boy, if I could remember the names of these particles I would have been a botanist.'"[2]

The particle explosion dismayed many physicists who hoped that their new atom smashers would reveal nature's underlying simplicity. "Instead we were beginning to count hundreds of particles," Lederman said. "It was a whole attic full of discoveries that came in so fast that we didn't know what to do with them. Except do what the botanists do, which is just classify them, organize them, and look for patterns."

Murray Gell-Mann was the best pattern finder of them all. Born in New York City in 1929, he was, to understate it, a bright child. By age 15 he entered Yale, and then he earned his Ph.D. at the Massachusetts Institute of Technology (MIT) by age 21. After a short stay at the Institute for Advanced Study in Princeton, Gell-Mann spent a few years at the University of Chicago in the early fifties, where he took the first giant step toward clearing up the particle muddle by inventing the idea of "strangeness."[3]

It was unusual in those days to give a new physics concept such a whimsical name. But Gell-Mann had precisely the right idea. After all, many of the new subatomic particles popping out of the atom smashers *were* strange; the V-zero-2 being an example. They weren't like the other particles that had become familiar by then, like protons and neutrons. The new particles differed somehow, in ways then dimly understood, from other particles. It seemed perfectly natural to describe them as *strange*, meaning they possessed a property called *strangeness*.

Strangeness was more than just a clever word, of course—it was a number. The strange new particles could be assigned a "strangeness number" that helped make sense of the situation. Typically the new particles appeared in pairs; if one of the particles had a strangeness number of 1, the other must have a strangeness value of –1, so that the total strangeness remained zero.

Strangeness was an original and fruitful idea. It eventually led Gell-Mann (who had moved on to the California Institute of Technology, known as Caltech) to the next step in understanding the basic units of matter—a pattern of particle properties that he called the Eightfold Way.[4]

In essence, Gell-Mann showed how to organize the species of the particle zoo (by now, nearly 100 particles had been discovered) into groups of 8 (or in some cases, 10). Amazingly, the properties of the particles could be described by the obscure (at the time) mathematical notion called group theory.[5] A family of 8 related particles, in Gell-Mann's scheme, corresponded to what a mathematician would have called a "group of eight." That's why Gell-Mann called his scheme the Eightfold Way.

Of course, he did not mean to imply that any eastern mysticism was involved in particle physics. ("I meant it as a joke!" he once exclaimed.)[6] His insights were much more in the spirit of the ancient Greeks, who two and a half millennia earlier had conceived of ultimate basic particles they called atoms, their word for "uncuttable."

INSIDE THE ATOM

In a way, the Greek idea of atoms represents one of the earliest great prediscoveries, the imagining of something to be discovered only much later. But the Greek concept was not so clear by the standards of modern physics. For a long time there was a lot of confusion about what the Greeks' idea of an atom really meant.

"They had two different concepts contained in the notion of an atom," Gell-Mann remarked over lunch when I interviewed him in Santa Fe in 1997. "One is the smallest unit of some substance, and the other was that it was—as the name indicates—uncuttable. But those turned out not to be the same thing."

Today's atoms are, in fact, the smallest units of chemical elements. But these atoms are clearly cuttable, or at least smashable and splittable, as they are made of still smaller pieces. In the sixties, Gell-Mann was searching for the true uncuttables, the most basic of matter's building blocks.

His Eightfold Way revealed the mathematical formulas corresponding to groups of subatomic particles. Understanding that math allowed him to arrange the known particles in charts similar to Dmitri Mendeleyev's periodic table of the chemical elements. The idea of positioning particles in a table like Mendeleyev's had been expressed years earlier by the physicist Abraham Pais. "The search for ordering principles at this moment may indeed ultimately have to be likened to a chemist's attempt to build up the periodic system if he were given only a dozen odd elements," Pais wrote.[7] Gell-Mann, on the other hand, told me he wasn't explicitly setting out in Mendeleyev's footsteps.

"I was playing around with the particles," Gell-Mann said. "He was playing around with the elements. It was natural to make a comparison between them, although I think Mendeleyev's work was much more important."[8]

Anyway, just as Mendeleyev had used gaps in his table to predict the existence of undiscovered elements, Gell-Mann predicted that new particles should be found to fill in some of the empty slots in his Eightfold Way charts. The new particles were found with just the properties he anticipated. In fact, when experimenters at a 1962 conference reported the discovery of new particles fitting into his scheme, Gell-Mann realized immediately that yet another new particle must exist, which he called the omega-minus. The mass of the new particle, Gell-Mann asserted, should be 1,685 million electron volts (MeV), and its strangeness number should be –3.[9]

Nicholas Samios, of Brookhaven, returned from the meeting determined to find Gell-Mann's new particle. It was not an easy experiment to do. But by December 1963, his team had succeeded in recording the particle collisions that Gell-Mann suggested might produce his omega-minus. In February 1964, the Brookhaven team published its report, announcing the discovery of a new particle with a mass of 1,686 MeV—give or take 12. The strangeness number: –3. Gell-Mann had been right on the money. The Brookhaven experimenters concluded that they were "justified in identifying" the new particle "with the sought-for omega-minus."[10]

It was one thing, though, to predict more particles that fit into established groups. It was something else again to imagine an entirely new sort of particle never previously encountered. But that's what Gell-Mann did next.

Doing so proved difficult, though. The math told him that the heavy particles in nature (known as the hadrons) could be built from a set of three basic particles. But this picture was blurred by a problem with electric charge. The math required the building-block particles to have electrical charges only one-third or two-thirds of the smallest unit of charge believed possible (the charge of an electron or proton). No experiment had ever produced a particle with fractional charge.

"I ignored the fractional charge possibility—it seemed so crazy," Gell-Mann told me. But at one point—in 1963—another physicist, Robert Serber, asked Gell-Mann why he hadn't used the triple-particle approach. Gell-Mann said that he had tried.

"I drew on a napkin a picture showing him the equation, showing him the charges would be fractional," Gell-Mann recalled. That seemed to satisfy Serber, but Gell-Mann pondered the three-particle approach some more. "During that day and the next day, I thought about it and I decided, well, maybe they don't come out, maybe they're trapped inside" the protons and neutrons and other hadrons.

So he pursued the idea of trapped, fractionally charged particles, to which he gave the name he now uses on his license plate: quarks. He referred to the quarks as "mathematical" or "fictitious," meaning, he said, that they would always be trapped inside a bigger particle and would therefore never be seen alone.

"That was the defining moment . . . after Bob Serber asked that question," Gell-Mann said. "I thought that maybe these things can't come out and therefore there's no problem with experiment."[11]

True enough, there was no problem with experiment then, and there hasn't been since. No compelling evidence has ever materialized for the existence of a free fractionally charged particle. Quarks are indeed trapped. Experimental evidence did come soon, though, for the reality of quarks trapped within other particles. Using the Stanford atom smasher, physicists in the late 1960s fired electrons into protons and found curious deflections of the electrons' paths— just the kind of deflections, in fact, to be expected if a proton was made up of smaller particles. It took a while for the physics community to reach a consensus that the proton's parts were the same as Gell-Mann's quarks. But by the 1980s, nobody seriously doubted it, and in 1990, Stanford physicist Richard Taylor, with MIT colleagues Henry Kendall and Jerome Friedman, won the Nobel Prize for their discovery. Gell-Mann's prediscovery of quarks had been verified.

In his original math, all the hadrons could be made from only three kinds of quarks—the up quark, designated by u, the down quark, symbolized by d, and a third quark abbreviated with the letter s—not for sideways, but for strange—the quark conferring the property that Gell-Mann had years earlier called strangeness. Protons and neutrons, being "ordinary particles," contained no strange quarks, just ups and downs. A proton contains two ups and a down; a neutron possesses two downs and an up. Thus a proton could be abbreviated as uud, a neutron as *udd*. Particles exhibiting strangeness contain combinations that include an *s*, or strange, quark.

Nowadays physicists recognize six quarks—the up, down, and strange quarks identified by Gell-Mann, and the charm, bottom, and top quarks discovered years later.

THE STRANGE QUARK STRIKES BACK

For many years after Gell-Mann invented them, nobody worried very much about strange quarks. After all, the oddball particles containing strange quarks didn't have a very prominent role in real life. The "strange" particles were created only under unusual conditions, and they didn't live very long. They were kind of like termites, unseen deep within the walls of nuclear science. But they were about to chew their way out.

In the early 1980s, a young physicist injected a strange new plot twist into the quark story. Edward Witten, at the time at Princeton University, was about to become the Murray Gell-Mann of his generation—a creative and critical thinker and the intellectual leader of an entire community of physicists.

It is fascinating to hear other physicists marvel at Witten's brilliance. I've attended numerous lectures where a speaker expressed awe at some insight Witten had provided to illuminate an important issue. One such comment came from Willy Fischler, a physicist at the University of Texas, when discussing a peculiar point that Witten had clarified involving string theory. Fischler admitted that he had no clue to how Witten had arrived at his conclusion. "I was not in his brain," Fischler said, "so I don't know."[12]

In his best-selling book *The Elegant Universe*, the theoretical physicist Brian Greene implies that Witten might be the greatest physicist who ever lived. I asked Greene if he really believed it. "I didn't say that," he said. "I said 'some would say' that he's the greatest physicist of all time."

"So do you agree?" I cross-examined.

"I think he's phenomenal," was Greene's evasive reply.[13]

In any event, when Edward Witten talks, other physicists listen. As they did in 1984, when he wrote a paper examining the idea that quarks need not always be confined in protons and neutrons. Under extreme conditions—say, within the superdense matter of a neutron star—perhaps quarks could arrange themselves differently, forming "quark matter."

At first glance, it seems like a far-fetched idea. We (and most of the matter we know about) are made of protons and neutrons (in this context, it is OK to ignore electrons). That must be because quarks like to stick together. Free-swimming quarks have energy to spare, a situation that physicists refer to as "unstable." Everything in nature seeks a condition where the energy required to maintain it is minimal. Sooner or later, energetic objects relax to the lowest-energy, stable state. Rocks roll down to the valleys below, living things die, and quarks coagulate to make protons and neutrons.

But suppose there is some more stable, lower-energy condition that quarks can find. In that case they might not need protons and neutrons. As early as 1971, Arnold Bodmer, of the University of Illinois, pointed out that up and down quarks might be stable if enough strange quarks joined them. With roughly equal numbers of strange, up, and down quarks, dense matter (inside stars, say) might actually prefer to stay in the form of a cluster of quarks, without congealing into protons and neutrons. Thus was born the novel idea of "strange quark matter"—either "strange matter" or "quark matter" for short.

Nobody paid much attention to Bodmer, although a few other physicists played around with the idea. But it became a hot topic in 1984, when Witten analyzed the strange situation.

Witten suspected that quark matter might solve a major astronomical mystery—the identity of the invisible "dark matter" that lurked throughout the cosmos, betraying its presence only by exerting gravitational effects on visible matter.

It seemed then (and still does) that the amount of dark matter is oddly similar to the amount of ordinary matter in the universe. Not exactly the same amount, but maybe seven times as much—to astrophysicists, not much of a difference. Maybe, Witten realized, dark matter was similar in quantity to ordinary matter because it was made of the same stuff (quarks), just arranged in a different way—in the form of quark matter, rather than as protons and neutrons. In the moments following the explosive big bang that gave birth to the universe, he suspected, the most stable arrangement of up, down, and strange quarks would have been within lumps of quark matter.

"I got the idea that if quark matter was stable at zero pressure, . . . perhaps the big bang was a good place to make it," he told me many years later.[14] If so, it raised the possibility that today, chunks of strange quark matter might be hanging out in the universe, accounting for the unseen dark matter that astronomers can't identify.

Witten realized that entertaining the idea of stable strange quark matter required a heavy dose of speculation. "The odds are against it," he acknowledged. Strange matter might be stable under astronomically high pressures, but in the zero-pressure environment of empty space it's not a good bet. But then again, long shots sometimes win.

"It's just barely possible that strange matter is stable even at zero pressure," Witten said. "Very hard to make, but stable once you make it." So Witten explored the notion of making strange quark matter in the big bang. Alas, he ran into a severe problem.

"There's really a very good objection . . . which comes up right away, which is that, unfortunately, the big bang was hot," he said. "If you accept at face value that the big bang was hot, it's really almost impossible to make quark matter."[15]

At high temperatures, you don't get lumps of strange quark matter, you get quark soup (or quark-gluon plasma). When it cools it wants to congeal into protons and neutrons, the process known to physicists as the quark-hadron phase transition (since protons and neutrons are hadrons).

In any event, the prospect that the dark matter is strange matter seems dim. Yet the idea of strange matter remains alive. It may not like the heat, but it loves pressure. Perhaps there are pressure cookers somewhere in the universe in which strange quark matter could thrive. One obvious possibility would be in the middle of a neutron star, where pressures dwarf anything ever encountered on Earth.

"We do have good reasons to believe that under sufficiently high pressure, quark matter would be stable," Witten told me. "But it's very difficult to estimate the pressure. . . . We don't know whether at the center of a neutron star the pressure is big enough. It's possible that it is, and that neutron stars are mostly strange matter. But it's also possible that even at the center of a neutron star you don't have quark matter."[16]

But if you did, then the possibility does exist that some strange matter might escape its neutron star prison—and if so, chunks of strange matter might indeed be found flying through space.

"If it's stable at zero pressure you could make it in neutron stars and it's conceivable that some catastrophes involving neutron stars would eject some into the universe," says Witten. "That's the best chance I can see."

So there is some small chance that some lumps of strange quark matter might be zipping through the universe, perhaps headed this way. Perhaps it would be worth looking for. "But only," says Vic Teplitz, "if you already have tenure."

THE SEISMIC TELESCOPE

Teplitz is a physicist who spends most of his time in an airplane, shuttling back and forth between Maryland, where he lives, and Dallas, where he works.[17] He came to Texas in 1990, eager to participate in the century's grandest physics experiment, the Superconducting Super Collider (SSC). In 1988 the government selected a site south of Dallas, surrounding the quaint town of Waxahachie, as the future center of high-energy physics. There physicists planned to oversee the construction of a 54-mile-around racetrack where protons would smash into each other with more energy than any atom smasher had ever before generated.

Five years later, not yet half-finished, the SSC was dead, killed by political bickering in Congress. In the meantime, though, Teplitz had revitalized the physics department at Southern Methodist University (SMU), a school more famous for football than science, tucked into the community of Highland Park, an "endoburb" entirely surrounded by the rest of Dallas.

It's not that SMU was without significant science, though. One of its most prominent programs involved seismology, where a nationally recognized expert named Eugene Herrin worked on problems like figuring out how to tell earthquakes from underground nuclear explosions.

One day Teplitz called me about a column I'd written for the *Dallas Morning News'* Monday science section. He was ecstatic that I'd written about a paper in *Physical Review Letters* describing strange quark matter. Some physicists in England and France had speculated that nuggets of strange matter might have survived the big bang and could be floating through space today. Apparently Herrin read the column, and it inspired him to call Teplitz to discuss a project Teplitz had been pushing Herrin to pursue. Teplitz wanted to build a seismic telescope—to search for those strange quark nuggets.

Since he had tenure, Teplitz figured he was free to pursue strange matter, but he needed the help of Herrin's seismic expertise to "build" a seismic telescope that might reveal strange quark visitors from space impinging on our planet. Basically, the seismic telescope he envisioned consisted primarily of the Earth itself.

As the sun and solar system sweep through the galaxy, Teplitz explained to me, the Earth might encounter quark nuggets in its path. A typical nugget might weigh 4 tons yet be smaller than a millimeter across. Traveling at 150 miles a second, it would shoot through Earth like a bullet through butter.[18]

The Earth would shudder at such an insult, with ripples from the passing nugget shaking the Earth's interior much like an earthquake or nuclear explosion. But an earthquake or explosion sends its waves outward from a single spot. The quark nugget's vibrations would emanate from all along its path through the planet. It should therefore be feasible to analyze readings by seismic recording stations around the globe to find strange patterns of signals from strange quark nuggets.

Fortunately, the U.S. Geological Survey collects seismic signal reports from stations around the world. For the years 1981 through 1993, about 9 million such signals are on record. And at least 2 million of those signals cannot be connected to known earthquakes. So with a little luck, and some superior computer programming skills, the SMU scientists might be able to analyze the mystery signals and find a pattern in which seven different stations recorded waves at the right time intervals to match the pattern expected of a quark nugget.

It turns out that programming a computer to analyze all that data and spot just the right combination of seismic tremors is a far from trivial task. And seeking strange quark matter is hard to justify as a full-time job. So merely preparing the computer has taken years, and analyzing all the data could take years longer.

"Nobody's been looking for strange quark nuggets passing through the earth, so things haven't been set up in an optimum way," Teplitz told me. And even in a best-case scenario, out of the 2 million or so events to sift through, there are probably no more than a handful of nuggets.

Of course, there might also be no nuggets at all. But Teplitz and Herrin are not alone in searching for them. Many experimenters have

expended considerable effort seeking signs of strange matter in mete-
orites, moon rocks, and the debris of atom-smasher collisions. So far,
no luck.

There are other attempts, though, to detect strange quark
nuggets striking the Earth. If the nuggets are a lot smaller than the
millimeter-sized lumps considered by Teplitz, they might not shoot
through the Earth but could be stopped by the atmosphere, behaving a lot like ordinary cosmic rays.

Shibaji Banerjee and colleagues at the Bose Institute, in Calcutta,
India, even suggest that smaller nuggets, or strangelets, may already
have been detected. A giveaway feature of a moderate-sized
strangelet is an unusually small electric charge compared to the mass
of the particle. An ordinary heavy atomic nucleus carries an electric
charge not too much less than half its mass. A typical uranium
nucleus, for example, has a mass of 238 (the sum of its protons and
neutrons) and a charge of 92 (the number of protons). But cosmic ray
detectors have on occasion reported signs of unusual nucleus-like
fragments, larger than uranium, but with a much smaller electrical
charge—as low as 14. Banerjee and friends have calculated that such
strange charges could in fact be explained by strange nuggets just
massive enough to slice through the Earth's atmosphere down to the
altitude of mountaintops.[19]

So there's hope that further analysis of cosmic rays may someday
offer evidence of strange matter. Still, it may be that strange nuggets
will remain forever strangers to the Earth. In that case, the best remaining bet for finding strange matter will be in space, based on the
hope that strange matter lurks inside neutron stars—or more precisely, what appear to be neutron stars. For it may be that neutron
stars are really made almost entirely of strange quark matter and
would therefore better be known as "strange" stars.

STRANGE STARS

The ultradense balls of nuclear matter known as neutron stars were discovered in 1967 by the British astronomer Jocelyn Bell, fulfilling a prediscovery from decades earlier. In 1932, the Russian physicist Lev Landau, inspired by the recent discovery of the neutron, suggested that stars made of neutrons might lurk in space. Two years later, Fritz Zwicky and Walter Baade proposed that the transformation of an ordinary star into a neutron star might underlie the phenomenon of exploding stars known as supernovas.

But in 1967, nobody realized at first that Bell's discovery was a neutron star, because it showed itself in an odd way—in the form of a pulsing radio signal, much like the beacon of a lighthouse. Soon the astronomer Thomas Gold realized that the pulses could be explained by the rotation of a neutron star surrounded by a magnetic field. Neutron stars emitting such signals became known as pulsars.

As their name implies, neutron stars, presumably, are made of neutrons. "Neutron matter" would be extremely dense, of course—something like a trillion times as dense as water, denser the deeper you go. In its earlier life as a shining star, a neutron star would have been more massive but less dense. Death comes to such a massive shining star when it has burned up all its nuclear fuel. It stops shining. And that means no pressure emanates from within to keep the star's mass from collapsing inward. The star's solution to this problem is to explode. As the star collapses upon itself, its inner core (made of iron) shrinks from a few thousand miles wide to a mere 20 miles across in about a second. Much like a compressed tennis ball, the iron core then rebounds, blasting the star's outer layers away (that's the supernova) the way a depressed trampoline would boost a gymnast into the air. All that's left is a dense core; that's the neutron star.

Basic physics suggests that this core must be even denser than an ordinary atomic nucleus (about 300 trillion times water's density). A chunk of such matter the size of a Fig Newton would weigh as much as 10 billion Nate Newtons (a very weighty football player). At such densities, a neutron star weighs a little more than the sun but would fit inside the interstate highway loop surrounding a big city.

Densities so high make the pressure in the middle of a neutron star enormous. And as you may remember from Chapter 1 (after all, you're still in Chapter 1), higher pressure raises the possibility that strange quark matter is stable. So suppose that somehow or other, conditions within a neutron star generated a bit of strange matter. The question is, could strange matter coexist inside the star with ordinary neutrons?

Maybe so, some analyses indicate. But the first appearance of strange matter might instead initiate a chain reaction. In minutes, the ordinary matter could all turn into strange matter, turning the neutron star into a "strange" star, releasing copious amounts of energy in the process—so much, in fact, that mysterious flashes of gamma rays from deep space might be produced in this way, some scientists suspect. If so, these gamma-ray bursts may actually be signals of strange matter in space.

Of course, there are other explanations for gamma-ray bursts. And in truth, nobody knows for sure whether strange matter *is* the most stable form of nuclear matter within neutron stars. There are even some indications that it isn't. In some pulsars, for example, the pulsing beams exhibit temporary hiccups, or glitches, indicating that the star's spin rate changed momentarily. Strange matter shouldn't be able to do that, conventional wisdom holds. But Norman Glendenning, of the Lawrence Berkeley Laboratory in California, and colleagues from Germany say dismissing strange matter for that reason is premature. It is possible, their calculations show, that strange stars could have glitches.

Still, the best evidence that strange stars exist would be their direct detection, and there are possible signatures of strange stars that could be observed from Earth. For one thing, extremely rapid pulses—on the order of 2,000 times a second—would be evidence of strangeness. A star made of neutron matter could not spin that fast without breaking apart. And Teplitz and Herrin (with additional collaborators from Virginia and California) have suggested yet another way to detect strange stars—namely, by tuning in to the right radio channel.[20]

If they exist, strange-matter stars just might be sending a radio signal advertising their presence. Teplitz, Herrin, and colleagues calculate that strange stars could emit radio signals with a wavelength of about a millimeter or two, right in a popular range of frequencies studied by radio telescopes.

Emission of such radio signals would be expected because strange-matter stars should acquire a thin crust of ordinary matter, cushioned from the strange interior by a layer of electrons. It is possible that this crust would vibrate back and forth with respect to the center of the core. Since the crust would be an electrical conductor, its motion would cause the cushioning electrons to slosh back and forth, generating a radio wave.

"A detectable signal could be achieved for pulsars as far away as 15 kiloparsecs (about 300 million billion miles)," the scientists wrote in their 1997 paper. More than 100 candidate pulsars are known within that distance.

OK, it sounds like another long shot. But even if all the searches for strange matter fail, there might still be some payoffs. All the unaccounted-for signals from inside the Earth studied by the "seismic telescope" may not be caused by strange nuggets but by some other interesting and previously unknown geological phenomenon. Furthermore, failing to find strange matter in cosmic rays or neutron stars would in itself tell a lot about its properties. It may sound strange, but understanding more about strange matter could put scientists on a path to better appreciating the universe's simplicity. As John Wheeler, the physicist who named black holes, has often remarked, "We will first understand how simple the universe is when we recognize how strange it is."

RHIC? RELAX

Meanwhile, fear of universal destruction from RHIC has faded. But the prospect of deadly danger from strange quark nuggets was worth considering. Strange quark nuggets are not the sort of thing you would want for a pet rock.

If somebody did create strange matter, it might pose the same danger to Earth as it would to a neutron star. A chunk of such strange matter forming within a neutron star could begin "eating" the neutrons surrounding it, converting them into additional strange matter—ultimately digesting the entire neutron star. So a strange quark nugget, or strangelet, does pose a certain danger to anything around it. You see why some people got worried about RHIC. As the team of experts studying RHIC's dangers acknowledged, a strange quark nugget is not something you'd want to meet in a dark alley, especially if it carried a negative electrical charge.

"If such an object did exist and could be produced at RHIC," wrote the analysis team, "it would indeed be extremely dangerous."

A whole chain of unlikely events would have to happen to wipe out the Earth, though. To begin with, RHIC would somehow have to produce a negatively charged strangelet that stays around at least a hundred-millionth of a second. If that happened, the strangelet would quickly be captured by an atomic nucleus in the vicinity. Once inside the nucleus, the strangelet would swallow other nuclear particles, beginning to grow into an even larger strangelet. For a moment, the strangelet's negative charge would switch to positive. But then it would begin gulping down negatively charged electrons in the vicinity of the nucleus, reversing charge again, back to negative, growing bigger all the while. Then it would capture more nuclear particles, and the process would repeat itself. Soon the strangelet would be 100 times the size of an ordinary atomic nucleus.

That would be the time to watch out. The electric charge of the strange quarks would initiate the creation of electrons and their anti-matter counterparts, positively charged positrons. The positrons would stick to the strangelet, surrounding it with a cloud of positive charge. Any atom passing by would find its electrons annihilated by the positrons, and the naked nucleus remaining would be gobbled up by the strangelet.

"This process would continue," the RHIC experts wrote, "until all available material had been converted to strange matter. We know of no absolute barrier to the rapid growth of a dangerous strangelet."[21] In other words, once a strangelet starts growing, it doesn't stop. Pretty soon the entire Earth would be just one big fat strange nugget.

It would be a very strange way to destroy the planet. But while this strangelet scenario is scarier than *The Blair Witch Project*, it's even more implausible. The RHIC experts conclude that there's really nothing to worry about.

First, there is no real evidence that strange matter can exist long enough for this to happen, they pointed out. Second, other particle smashers had tried to make strange matter and failed. And if RHIC's collisions could make strange matter, so would have cosmic rays colliding with the moon. And the moon is still up there.

Besides, if RHIC did accidentally make a strangelet, it would almost certainly have a positive electrical charge, and only negatively charged strangelets are dangerous. In other words, worrying about strangelets is like fearing your dog'll bite you to death, when your dog has no teeth, and you don't even have a dog.

As it turned out, RHIC didn't get going on schedule, anyway; it was mid-2000 before it began smashing gold atoms together in earnest. By then Y2K fears had fizzled, along with any worries that

RHIC would destroy the universe. But the search for strange quark matter continued.

In fact, in April 2002 two teams of astronomers reported new evidence for strange quark stars, announcing at a NASA news conference that data on two supposed neutron stars pointed to the conclusion that they were made of strange quark matter. "Stars suggest a quark twist and a new kind of matter," proclaimed the headline in the *New York Times*.

And a day before the NASA news conference, Teplitz and Herrin submitted a paper for publication, reporting the possible "sighting" of two strange quark nuggets in their analysis of seismic data. So by the time you're reading this book, my promise in the preface that at least one of the strange matters will someday be discovered may already have come true.

In the meantime, Teplitz has been spending some time looking for other potential prediscoveries in a mirror. Or more precisely, in a mirror world.

MIRROR MATTER

2

From Dirac and Antimatter
to the "Mirror World"

A great deal of my work is just playing with equations and seeing what they give.

—Paul Dirac

A mirror macho is not a big tough guy who spends the day looking at his face in the mirror.

It's a macho—make that MACHO—made of matter from the mirror world, a hypothetical wonderland invented by physicists who didn't have enough subatomic particles to play with.

Those physicists will tell you that every type of particle in nature has a "mirror partner." If you looked into a mirror made of mirror matter, though, you might as well be a vampire, because you would see no reflection. In fact, you couldn't even see the mirror. Mirror matter is utterly invisible; it does not interact with light. Mirror matter can be detected only by its gravity.

35

If mirror matter is real, there's a lot more matter in the universe than astronomers can see. But guess what: astronomers already know that there is a lot more matter in the universe than they can see. They've been looking for it for years. Some of that "dark matter" lurks throughout the outer regions (or "halos") of galaxies in the form of massive compact objects the size of small stars. Those *massive* compact *halo objects* are what scientists call MACHOs.

Astronomers have actually detected a handful of these MACHOs but aren't really sure exactly what they are. Most MACHOs seem to be roughly half the mass of the sun, which would be the right size for burnt-out stars known as white dwarfs. But there's no way the galaxy could have made enough white dwarfs to account for very much of the halo matter. Brown dwarfs are too small to be the MACHOs. Red dwarfs have also been ruled out. So MACHOs are a mystery. But some scientists think that solving it is as simple as looking in a mirror.

It is, of course, a mathematical mirror. Certain formulas suggest that every type of subatomic particle known to science is something like one of a pair of gloves—either left- or right-handed. The opposite glove should be out there, somewhere. Of course, there is no compelling reason for believing this possibility other than that the math suggests that it might be true. But scientists have learned through experience to respect what mathematics tells them, and no one taught that lesson more dramatically than Paul Adrien Maurice Dirac.

THE SILENT TYPE

For Dirac, numbers always spoke louder than words. He was no macho man, but rather one of the twentieth century's shiest and quietest physicists. He was unquestionably brilliant but handicapped by a harsh childhood, leaving him without a normal repertoire of human interaction skills. He remarked once that he had been brought up "without any social contact." His father, a stern high-school French teacher, insisted that while at home Paul speak only in French. Paul lacked confidence in his French and resolved this inner conflict by rarely speaking at all. "Since I found I couldn't express myself in French, it was better for me to stay silent than to talk in English," he recalled.[1]

Terseness marked Dirac's style for the rest of his life. A colleague once remarked that Dirac was suspected of knowing only three phrases: "Yes, no, and I don't know." When he did speak, his comments were always pointed. George Gamow recalled that after a lecture in Canada, Dirac asked for questions.

"I do not understand how you derived this formula," a professor in the audience said. "This is a statement and not a question," Dirac responded. "Next question, please."[2]

Born in 1902 in Bristol, England, Dirac earned a degree in electrical engineering at the university there but couldn't find a good job. He'd shown by then that he was a whiz at math, though, so Bristol offered him funding to spend two extra years as a math student. Afterwards Dirac went to Cambridge, where he mixed math with theoretical physics, and he found himself smack in the middle of the biggest scientific revolution since Newton. As the first phase of that revolution ended, Dirac produced what many regard as the greatest example of prediscovery in the history of physics: antimatter.

Dirac's prediscovery grew from his efforts to understand the meaning of the math behind quantum mechanics, a field of study that didn't even exist when he entered Cambridge in 1923. In those days physicists struggled with trying to understand the "old quantum theory," the mathematical predecessor of quantum mechanics. The centerpiece of the old version was Niels Bohr's quantum theory of the atom, published in 1913. But Dirac heard about it for the first time only after entering Cambridge a decade later.

In his 1913 papers, Bohr used the then-novel quantum ideas to explain the structure of the simplest atom in nature, hydrogen. Like other atoms, hydrogen emits specific colors of light when heated up or otherwise energized. Various colors identify atoms like the flags of different nations. By recording the colors coming from a gas, you can deduce precisely what atoms it contains. (In this way, scientists were able to discover the element helium in the sun, many years before it was found on Earth, and can tell you what chemicals have been cooked up in distant stars.)

Scientists had long known about the colors and presumed that they had something to do with the way atoms were put together. But nobody knew how. A big clue emerged in 1911, though, when Ernest Rutherford, a New Zealander working then in England at Manchester, figured out that atoms consisted mostly of empty space.

Rutherford had instructed his assistants to fire alpha particles (subatomic bullets emitted by some radioactive atoms) at a thin gold foil surrounded by phosphorescent detectors. As expected, most of the alpha particles sailed through the foil—it was too thin to stop a fast-charging subatomic particle. But a few of the alpha particles were diverted far from their path, and some bounced almost straight back. Rutherford eventually deduced that the foil must contain some tiny, dense bits of matter, kind of like cherry pits, that unlucky alpha particles smashed into. With some simple calculations Rutherford showed that a very small, very dense nucleus resided in every atom's core, carrying a positive electrical charge. Negatively charged electrons, lightweights compared to the nucleus, should be speeding about an atom's outer edges, perhaps like planets orbiting the sun. An atom of hydrogen, for example, possessed a nucleus consisting of a single particle—a proton—with a single electron for its "planet."

While Rutherford's nucleus model offered a clue, it also posed a problem. An electron in orbit, with an electrical charge, should spit out light of some sort all the time. After all, light is just a form of electromagnetic radiation, and a charged particle changing direction should emit such radiation in the process, as James Clerk Maxwell had demonstrated in the nineteenth century. In fact, a simple calculation suggested that an electron would emit its energy so rapidly that it would spiral into the hydrogen nucleus in a fraction of a second, making hydrogen atoms rather too short-lived to be good for anything. Clearly hydrogen (and most other atoms) lived a lot longer than that. Somehow atoms remained in a stable, or "stationary," state, emitting radiation only when provoked. Bohr, a Dane who had gone to Manchester for postdoctoral study with Rutherford, tackled this problem in 1913.

Among the theoretical physicists of the twentieth century, Bohr ranked second only to Einstein in intellectual power and influence. But unlike most theorists, who guide themselves through nature by interpreting maps written in mathematics, Bohr sought a deeper physical understanding. He wanted to know what was really going on—a difficult task when dealing in the invisible realm of the atom's interior. While working with Rutherford, Bohr began to form his mental picture of the atomic blueprint for hydrogen. But at first he did not see how the pattern of colors that hydrogen emitted could help him.

"One thought that this is marvelous, but it is not possible to make progress there," Bohr said years later, recalling his early efforts. "Just as if you have the wing of a butterfly, then certainly it is very regular with the colors and so on, but nobody thought that one could get the basis of biology from the coloring of the wing of a butterfly."[3]

But then Bohr was told of a formula for the frequencies of light in the hydrogen spectrum, devised in 1885 by a Swiss mathematician named Johann Jakob Balmer. Balmer had no idea why his formula worked; he had merely discovered certain numerical tricks by which the frequency of some individual lines in the spectrum could be related to the frequencies of other lines. Balmer's simple formula provided Bohr with a Eureka Moment.

"As soon as I saw Balmer's formula," Bohr remembered, "the whole thing was immediately clear to me."[4] Specifically, it became clear to Bohr that he could explain the hydrogen atom by borrowing the still-young quantum theory, delivered to the world in 1900 by the German physicist Max Planck. Planck, studying the colors of light emitted from a hot cavity (sort of like an oven) concluded that energy gets absorbed or emitted in chunks, which he called quanta. Bohr showed how Planck's idea could explain the pattern of colors coming from hydrogen. An electron could swallow a quantum of light (or other form of electromagnetic radiation) and jump into a higher, more energetic orbit. At some later time the electron could fall back into the closer, lower-energy orbit, emitting a definite color of light in the process.

Hydrogen's colors depended on the size of the gaps between different possible orbits. An electron falling from a high orbit to one much lower would give off higher-energy radiation—ultraviolet light or maybe even an X ray. A jump between closer-spaced orbits would emit lower-energy light. Of course, the key to this picture was that only certain orbits are allowed—the electron could not hang out anywhere in between its permitted flight zones. Different atoms permitted different orbits, which was why each atom gave off its own set of colors.

When introduced to Bohr's quantum atom in Cambridge, Dirac was instantly entranced. "I remember what a surprise it was to me when I first learned about the Bohr theory," Dirac recalled in a 1975 lecture. "I still remember very well how strongly I was impressed. . . . It is really the most unexpected, the most surprising thing that such a radical departure from the laws of Newton should be successful."[5]

Bohr's model succeeded spectacularly in explaining the colors coming from hydrogen. But it didn't work so well for helium—or for any other atom, either. Somehow the simplicity of the hydrogen atom, with only one electron, made it possible to compute the correct energy levels for the electron orbits using Bohr's theory. All other atoms contained more than one electron, and Bohr's theory broke down under the additional load.

"That was the situation," Dirac recalled, "when I first started research on atomic theory."

QUANTUM BREAKTHROUGHS

The year was now 1925, and Werner Heisenberg, one of Bohr's protégés, suffered an attack of hay fever and ventured therefore to a remote island in the North Sea so he could breathe while he worked. Without the distractions of city sounds or airborne allergens, Heisenberg produced some strange looking math that seemed to solve the problem of multiple electrons. He decided to forget about the electron orbits—which couldn't really be observed anyway—and focus on quantities that could be measured, such as the color (or frequency) differences between "orbits." He worked out a way to describe frequencies using an array of numbers, called a matrix (although he didn't know it was called a matrix at the time[6]). With help from his professor at the University of Göttingen, Max Born, and the mathematician Pascual Jordan, Heisenberg's breakthrough led to the first formulation of the mathematical framework nowadays known as quantum mechanics.

When Dirac received an advance copy of Heisenberg's first paper, in September 1925, he studied it for a few weeks, and then was struck during a Sunday walk with a mathematical insight, enabling him to reconstruct Heisenberg's findings a little more elegantly. Soon, in 1926, the Austrian physicist Erwin Schrödinger offered yet another approach, known as wave mechanics, describing the electron orbits as closed waves encircling a nucleus. After a short period of consternation, it became clear to everybody that all these approaches ended up being mathematically equal. Dirac, Schrödinger and Heisenberg, Born and Jordan are generally regarded as the originators of quantum mechanics.

With the breakthroughs of 1925 and 1926, the job was still not done, however. Dirac realized, more so than the other players in the quantum game, that Einstein's special relativity theory had to be inserted into the action. Schrödinger had at first attempted to derive a form of his wave equation that incorporated relativity. But when he calculated the results, he found answers that did not agree very well with the best experimental measurements of the day. So he published the nonrelativistic version instead. As it turned out, the experiments had been inaccurate. Schrödinger blew it.

"Schrödinger lacked courage to publish an equation that gave results in disagreement with observation," Dirac commented later.[7] But as it turned out, Dirac was soon to exhibit a certain lack of courage as well.

"There was a real difficulty in making the quantum mechanics agree with relativity," Dirac recalled half a century later. "That difficulty bothered me very much at the time, but it did not seem to bother other physicists, for some reason which I am not very clear about."[8]

Dirac was, however, not the only physicist to pursue a relativistic description of the electron. Oskar Klein, working at Bohr's physics institute in Copenhagen, had already produced an equation describing the electron that incorporated the math of Einstein's relativity. Dirac was not satisfied with Klein's version, though. "I was worrying over this point for some months," Dirac reported.[9] Ultimately, he found a new equation—more in tune with the basic principles of quantum mechanics than Klein's. Much to Dirac's amazement, the equation held within it the notion of electron spin, the property at the root of magnetism. Electron spin itself had just been discovered.

On the other hand, the new equation posed a somewhat thorny problem. It permitted electrons to possess negative energy.

NEGATIVITY

Now, negative energy is one of those amazing concepts that make quantum physics so interesting. At first glance, it's absurd, but on closer examination, it's deeply intriguing. How can an electron have negative energy? You would think it either has no energy or some energy, and that less than no energy is nonsense. But think a little more. It is possible to have less than no money, for instance, if you spend too much on credit. You could earn a lot of money yet still not have any, as all your earnings went straight to your creditors. There is certainly something real about being in debt.

And in a not too dissimilar way, there is something real about negative energy.

For Dirac, it was simply a matter of listening to what the math had to say. In this case, the math was speaking in the language of square roots. So if you've seen the movie *Stand and Deliver*, you know basically all you need to know to understand Dirac's prediction of the existence of antimatter.

In algebra, negative numbers come into play all the time, just as they do sometimes in bank statements. If you subtract a big number from a smaller number, the result is a negative number. If you want to multiply negative numbers, it gets a little more complicated. In *Stand and Deliver* (1987), math teacher Jaime Escalante, played by Edward James Olmos, drills a basic fact of algebra into his students' heads by forcing them to repeat, over and over, that "a negative times a negative equals a positive." So, for instance, a negative 2 times a negative 2 equals a positive 4.

It's a good thing to know for passing algebra tests, and it also turns out to be important in physics, as Dirac emphasized, especially when dealing with square roots. The square root of a number is simply some other number that, when multiplied by itself, yields the original number: 2 times 2 is 4, so 2 is the square root of 4. But if you have an equation with a square root in it, you must not forget that there are usually two solutions—negative and positive. Sure, 2 is the square root of 4. But so is negative 2, because negative 2 times negative 2 is also equal to 4. A negative times a negative equals a positive.

Now in Dirac's math, the equation giving the energy of an electron had a square root in it. (It's a consequence of applying Einstein's special theory of relativity, which includes the square of the speed of light in the energy formula.)[10] Dirac saw no escape: electrons could possess either negative or positive energy.

Did that matter? Well, in the old world of Newtonian physics, it wouldn't have. You could show (or at least Dirac could show) that a particle starting out in life with negative energy could never attain a state of positive energy. And a particle starting out life with positive energy could never slow down so much that it had negative energy—could never go into energy debt, so to speak. So the possibility of negative energy could be safely ignored.

But in the new era of quantum mechanics and Einstein's relativity, the situation changed significantly. Under the quantum rules, electrons could make quantum jumps; a positive energy electron might jump to a negative state, or vice versa. At least, Dirac reasoned, that possibility should not be ignored, even though all other physicists of the day were in fact ignoring it.

So when Dirac published his paper containing his electron equation (known evermore as the Dirac equation) in 1928, he pointed out the problem with negative energies, suggesting that maybe they had something to do with particles carrying an electric charge opposite the electron's. He was on the verge of anticipating the existence of antimatter, but hesitated.

CHICKENING OUT

At the time, science knew of only two basic particles in nature: protons and electrons. Everybody believed that the atomic nucleus contained both protons and electrons (but always more protons than electrons, so the nucleus would retain a positive charge). It seemed like a very neat way to make the world—two kinds of particles, co-operating to make atoms that could do all the wonderful things that nature does. So Dirac began to suspect that maybe his electron theory could also explain protons—if you thought of physicists (and all other people) as being something like fish in water.

Presumably (although I'm not sure fish would agree), a fish has no sense of living in an ocean of H_2O. At least a fish would give no more thought to the water around it than people ordinarily do to the atmosphere. In a similar way, Dirac reasoned, scientists would never notice a uniform sea of particles if it engulfed all of us all the time. And that's just what he thought was going on with the negative-energy electrons.

All particles of matter are ultimately couch potatoes—as lazy as they can be. In physics terms, that means seeking the lowest possible state of energy. Like the rock on top of a mountain that would like to roll downhill, electrons in an atom are always trying to fall to the lowest energy orbit whenever a spot is available.

But why only if an open spot is available? Why don't all electrons fall all the way to the nucleus, as low an energy as they could get? Because they aren't allowed to, prohibited by a declaration from the Austrian physicist Wolfgang Pauli. No two electrons could occupy the same energy state, Pauli had determined in the midst of the quantum mechanics revolution. His "exclusion principle" served a valuable guidance role in the efforts to understand the electron.

Applying Pauli's principle to the problem, Dirac realized that the vacuum of space could be filled up with negative-energy electrons. All the electrons in the universe would have sought negative energy levels to get as low an energy berth as they could find. Sooner or later, all the negative energy states would have been filled up, and Pauli's principle would have allowed none to enter after that. So the leftover electrons, forced to maintain positive energy, are the ones that scientists observe and that play important roles in daily life. The negative ones make up a smooth undetectable ocean in which people go about daily life unaware.

But even a fish occasionally notices something fishy about its ocean world, as when air bubbles gurgle by. To a fish, a bubble might seem like a small particle traveling through "space." In other words, the absence of water looks not like nothing, but like something. And so, Dirac reasoned, maybe every once in a while some jolt of energy kicked a negative-energy electron out of its ocean. We would then see a hole in the ocean that would look to us much like a bubble to a fish—a definite particle moving through space.

"Let us assume . . . that all the states of negative energy are occupied except perhaps a few of small velocity," Dirac wrote in a 1930 paper spelling out these ideas. "We shall have an infinite number of electrons in negative-energy states . . . but if their distribution is exactly uniform we should expect them to be completely unobservable. Only the small departures from exact uniformity, brought about by some of the negative-energy states being unoccupied, can we hope to observe."[11]

But what, precisely, would scientists observe? A particle of course, but what kind? Here Dirac was on the verge of prediscovering a new kind of matter, namely "antiparticles." But, a bit like Schrödinger, Dirac lacked courage. The bubble-particle, he knew, would appear to carry a positive charge, since a gap in a sea of undetected negative charge would be positive by comparison. Dirac, along with every other physicist, knew of only one particle in nature that carried a positive charge—the proton. So he suggested that the holes were protons. But protons were known to weigh nearly 2,000 times as much as an electron. Dirac could not explain how a hole the size of an electron could have a mass 2,000 times greater.

"When I first thought of this idea, it occurred to me that the mass would have to be the same as that of the electron because of the symmetry," Dirac remembered. "But I did not dare to put forward that idea, because it seemed to me that if this new kind of particle (having the same mass as the electron and an opposite charge) existed, it would certainly have been discovered by the experimenters."[12]

In other words, Dirac chickened out.

"That, of course, was really quite wrong of me," he admitted later. "It was just lack of boldness. I should have said in the first place that the 'hole' would have to have the same mass as the original electrons."

Soon the mathematician Hermann Weyl noted this discrepancy and argued that purely on mathematical grounds the mass of the hole would have to be the same as the mass of an electron. And then J. Robert Oppenheimer, later to become the father of the atomic bomb, ripped the proton idea to shreds in a 1930 paper published in *The Physical Review*.

"There are several grave difficulties which arise when one tries to maintain the suggestion that the protons are gaps of negative energy," Oppenheimer wrote.[13] For one thing, there are lots of protons around. If they are really "holes" in an electron sea, positive-energy electrons would constantly be falling into them and thereby disappearing (and the proton would disappear as well). In fact, Oppenheimer calculated, an ordinary electron should encounter a proton and disappear in about one ten-billionth of a second. Whereas in fact, electrons and protons happily coexist for much longer times than that. Therefore, Oppenheimer concluded, there are no holes. All the negative electron locations remain filled.

"Oppenheimer just said that there was some reason, which we do not understand, why the holes are never observed," Dirac recalled. In other words, Oppenheimer also chickened out.

By this time, though, Dirac was ready to accept the consequences of his own mathematical actions, and he wrote a paper explicitly predicting that the holes would be seen as new particles, positively charged, with precisely the mass of the electron. "A hole, if there were one, would be a new kind of particle, unknown to experimental physics, having the same mass and opposite charge of the electron," Dirac wrote in his new paper, which appeared in May of 1931.[14]

So even though he reached this conclusion rather timidly, Dirac nevertheless foresaw the discovery of an entirely new type of basic particle of matter. "I think Dirac's prediction of antiparticles is the most dramatic prediction in the history of science," the physicist Gordon Kane told me. "Nobody ever predicted new things like that before."[15]

BLIND OBSERVERS

Of course, Dirac had not really answered Oppenheimer's question: Why had nobody seen these "antielectrons"? Years later, though, Dirac offered an explanation: "The reason why the holes were not observed was simply that the experimental people had not looked for them in the right place, or if they had looked, they had not recognized what they saw."[16] Experimenters, Dirac proclaimed, "were prejudiced against new particles."[17]

In fact, the experimenters *had* seen the antielectrons, without realizing it. It was common in those days to use cloud chambers to study the particles known as cosmic rays that assaulted the Earth from outer space. Particles passing through such a chamber leave visible tracks as vapor condenses along the route that the particle takes. Add a magnet, and the path of any electrically charged particle will bend, depending on the particle's mass and amount of charge. Pictures of the chamber can then be studied to analyze the paths and identify the particles that made them.

Using this technique, Patrick M. S. Blackett, at Cambridge, actually detected some strange cosmic ray particle tracks that turned out to be Dirac's antielectrons. Dirac even told Blackett that's what they were. But Blackett remained unconvinced and failed to publish his data. Blackett chickened out, too.

A young American in California, on the other hand, followed the particle tracks where they led.

Carl Anderson, born in 1905 in New York City, headed west as a very young man (he was 7) and grew up in Los Angeles and then went to college at Caltech in Pasadena (despite the warnings of teachers who told him if he managed to get accepted, he'd probably just flunk out). By 1930, he had earned his Caltech Ph.D.

Anderson wanted to stay on at Caltech as a postdoc, but the school's president, the famous Robert Millikan, advised otherwise. Having all your degrees from the same school is a good reason to go somewhere else for a while, Millikan said, and he recommended that Anderson apply for a National Research Council fellowship to continue his studies elsewhere. So Anderson made plans to go to the University of Chicago and soon convinced himself that it was much the better opportunity.

His enthusiasm for Chicago at a peak, Anderson was then summoned to Millikan's office. Millikan had changed his mind. Famous for measuring the electrical charge on the electron, Millikan now wanted to gather good data on the energy of electrons in cosmic rays. He needed a postdoc with Anderson's expertise. Anderson, now eager to head east, protested. "I used all the arguments that he had previously made as to why I should not stay on at Caltech," he recalled.[18] But Millikan offered a strong counterargument—Anderson had not yet been granted the National Research Council fellowship, and Millikan was on the fellowship selection committee. Anderson's chances of getting the fellowship would be a lot better if he stayed another year at Caltech, Millikan mentioned.

So, in a sort of reverse serendipity, Anderson stayed and studied the electrons in cosmic rays. Some of those electrons behaved oddly. Their negative charge should have caused them all to follow a similar curved path in the cloud chamber's magnetic field. But the photographs showed some particles curving in the opposite direction. "Something new and mysterious must be occurring," Anderson concluded.

Maybe the particles curving the opposite way were protons— positively charged and therefore expected to curve in paths opposite to electrons. But further analysis showed that many of the particles were much too light to be protons. Perhaps, Anderson believed, they were electrons traveling upward rather than downward. That would explain the opposite curve. But Millikan didn't like that explanation, insisting that cosmic ray particles came down from above, not up from below.

So Anderson modified the experiment, putting a thin lead plate in the middle of the chamber to slow the particles down and thereby show whether they were traveling upwards or downwards. (Slower particles have less momentum and therefore resist the pull of the magnetic field less and curve more sharply. A particle track coming in from above the plate would curve more sharply below it.) When he inspected the new photographs, Anderson was doubly surprised. For one of the pictures showed a particle traveling upward—and turning in the direction opposite that of ordinary electrons. Yet its mass was obviously very close to the mass of an electron. It was, Anderson realized, an electron with a positive charge. He called it a "positive electron," a term he later shortened to positron, and published the report of its discovery in *Science* on September 9, 1932. It was, in fact, the discovery of Dirac's antielectron.

Curiously, though, Anderson did not seem very impressed by Dirac's anticipation of this discovery. In the account of the positron discovery in his autobiography, Anderson doesn't mention Dirac at all. In a lecture presented for Anderson at a 1980 conference (he was unable to attend himself), he explicitly discounts Dirac's contribution.

"It has often been stated in the literature that the discovery of the positron was a consequence of its theoretical prediction by Paul A. M. Dirac, but this is not true," Anderson declared. "The discovery of the positron was wholly accidental. Despite the fact that Dirac's relativistic theory of the electron was an excellent theory of the positron, and despite the fact that the existence of this theory was well known to nearly all physicists, including me, it played no part whatsoever in the discovery of the positron."[19]

Frankly, I find this attitude rather strange—if Anderson knew about the antiparticle theory, how could he not have been influenced by it? In any case, the fact remains that Dirac anticipated Anderson's discovery. By exploring the implications of squiggles on paper, Dirac had deduced the existence of something that nature had been concealing from the inquiring eyes of the observers.

SYMMETRY STRIKES AGAIN

As Dirac soon realized, it makes no sense to say that only electrons have antimatter counterparts. The same mathematical reasoning applies to every other sort of particle as well. So there must be antiprotons, for example, and there are—although they weren't actually observed until 1955. All the other particles discovered in later years have antiparticles, too (if you count the occasional odd case of certain particles, such as the photon, whose antiparticle is exactly the same as itself).[20]

Antimatter's existence provides nature with a nice example of symmetry. To physicists, symmetry is more than pretty wallpaper patterns or kaleidoscopic colors. It's a deep mathematical expression of constancy in nature. Describing nature with laws, expressed by unchanging math, requires a faith in something that remains constant beneath all the obvious changes in the world. Physicists express that concept in terms of symmetry. Symmetry allows change without change. Convert every particle in nature to its antiparticle, for example, and everything seems to remain the same. All the original laws of physics continue to apply with equal accuracy. Well, almost.

It turns out that keeping everything constant requires more than just giving every particle of antimatter an opposite electrical charge. At first glance, all natural processes would appear to happen in the same way in an antimatter world. But if you looked closely, you'd notice a subtle difference. Pictures of some processes would appear as though printed with the negative flipped upside down. To make them look like the original real world, you'd have to view them in a mirror.

In other words, when antimatter enters the picture, preserving the symmetries in the laws of physics requires switching left with right—converting the picture into its mirror image. In physicsspeak, that's called reversing the parity. (Parity is a fancy word for mirror symmetry.)

You might expect—and until the mid-1950s, every physicist would have agreed with your suspicion—that the universe in a mirror would look just like the universe always looks. Left and right would be switched, but the laws of nature would all work in just the same way. A baseball player would have to start running toward third base, then to second, then first, but would arrive back at home plate to score a run in the ordinary way, and otherwise the rules of baseball would stay the same.

In nature, things almost work that way, but not quite. Flipping parity leaves almost everything the same, but only almost. Somehow, the universe seems to know left from right.

MIRROR, MIRROR

Before the 1950s, nobody suspected that nature knew the difference between left and right. "There can be no doubt," Hermann Weyl once wrote, "that all natural laws are invariant with respect to an interchange of right and left."[21] It seemed pretty obvious that the laws of nature applied equally to Lefty Gomez and Bob Feller or Ted Williams and Joe DiMaggio. But in 1956, Chen Ning Yang and Tsung Dao Lee published an insightful paper in The Physical Review pointing out that the evidence for left-right symmetry was weak in cases where the nuclear force was also weak—specifically, in processes whereby subatomic particles disintegrated into other particles.

Soon careful experiments based on Yang and Lee's suggestions proved that left-right symmetry was indeed violated in weak nuclear interactions, such as the form of radioactivity known as beta decay. In one of those experiments, atoms of a radioactive form of cobalt (known as cobalt-60) were lined up in a magnetic field, so the nucleus of every cobalt atom would be spinning in the same direction. (To keep the nuclei lined up like that, the apparatus must be maintained at extremely low temperatures, close to absolute zero.) When a cobalt-60 nucleus decays, it spits out an electron, or beta particle. To test for conservation of parity, it is necessary to record the direction in which the electrons emerge from the cobalt nuclei (that is, whether they fly off in the same direction as the nuclei are spinning, or in the opposite direction). That trick is accomplished by using two detectors, one positioned to record electrons going one direction, the second placed in the path of electrons going the other direction. Sure enough, when Chien-Shiung Wu and a team of collaborators conducted the experiment at Columbia University late in 1956, they found that one detector recorded more electrons than the other, demonstrating the inequality of left and right in the beta decay of cobalt-60.

Of course, maybe there was just something funny about cobalt-60, and parity violation was not a general feature of weak nuclear interactions. But shortly after the Wu experiment, Leon Lederman and colleagues at Columbia tried another approach, using the subatomic particle known as the muon. Muons are unstable, giving off electrons when they decay, in another example of the weak nuclear interactions at work. Lederman and his collaborators realized that in a beam of muons, the electrons produced by muon decay would emerge more in one direction than the other if parity symmetry is violated. It was a technically challenging experiment, but it succeeded. Early one morning in January 1957, Lederman called Lee on the phone and announced, "Parity is dead."[22]

It's not overly dramatic to say that the physics world was shocked. "It was socko!" recalled Lederman.[23]

Reflecting on the situation, Yang and Lee had pointed out in their paper that there was a way to restore left-right symmetry to nature. Sort of the way Dirac enforced charge symmetry by requiring every known particle to have an oppositely charged antimatter counterpart, Yang and Lee proposed that every right-handed particle might have a left-handed counterpart, and vice versa. In other words, in addition to the antimatter world, there might also exist a mirror world.

"If such asymmetry is indeed found, the question could still be raised whether there could not exist corresponding elementary particles exhibiting opposite asymmetry such that in the broader sense there will still be over-all right-left symmetry," Yang and Lee wrote.[24]

For instance, they speculated, "normal" protons might all be from the "right-handed" world. For all we knew, corresponding left-hand world protons might exist as well but merely were exceedingly rare in our corner of the universe. Perhaps both right-handed and left-handed protons might interact with the same electromagnetic field, Yang and Lee suggested. But as other physicists remarked later, maybe the mirror particles did *not* interact electromagnetically—in which case a universe of mirror-image matter would be invisible. So in a certain sense, the cosmos would have something in common with vampires. In principle, all the basic particles of ordinary matter could have invisible twins in the mirror world, governed by mirror forces, perhaps forming mirror stars, mirror planets, and mirror people.

While such ideas are an offshoot of Yang and Lee's original suggestion, the math is much more sophisticated now. Strictly on the basis of those squiggles on paper, it now appears that nature does in fact allow a complete set of mirror particles. If the math is right, the only effect of mirror matter on ordinary matter would be via the force of gravity. People made of ordinary matter could neither see, feel, nor smell mirror matter. A normal-matter object encountering mirror matter would pass right through it.[25]

Mirror matter is truly one of the most fantastic not-yet-discovered ideas floating around the world of physics today. Yet it's one that many physicists take seriously. Starting with Yang and Lee's suspicion, several physicists have speculated about the existence of a mirror world. But like most prediscoveries, mirror matter will be a bigger deal if somebody actually finds it. Unfortunately, it's not the sort of thing where you can say you'll know it when you see it, because it can't be seen. So the possibility exists that mirror matter's presence has already been detected, but physicists just don't believe it yet.

I recall some excitement about mirror matter in the 1980s, but I didn't take it seriously until 1996, when Vic Teplitz told me about a paper he had written with Rabindra Mohapatra, a physicist at the University of Maryland. A year earlier, Mohapatra and a Russian physicist, Zurab Berezhiani of the Georgian Academy of Sciences in Tbilisi, Georgia, invoked mirror matter to explain one of the greatest mysteries in astronomy, having to do with how the sun shines.

Since the 1930s, physicists had understood that nuclear reactions deep in the sun produce the energy that makes it shine so brightly, supplying us with ample heat and light. Those nuclear reactions produce tiny particles called neutrinos (another example of prediscovery to be discussed in Chapter 4). Many neutrinos stream from the sun into space, some passing through the Earth. But scientists count too few neutrinos to account for the fact that the sun shines the way it is supposed to. Mohapatra and Berezhiani proposed that the solar neutrino mystery, and other neutrino oddities as well, could be explained by the existence of a "mirror" neutrino.

Teplitz explained to me that if mirror neutrinos existed, no doubt other mirror particles did also; in fact, the universe could be as full of mirror matter as it is of the matter we can see. Alas, he confessed, the prospect of parallel mirror civilizations turned out to be quite un-likely. If mirror matter really consisted of identical counterparts to ordinary matter, the opening instants of the universe would have cooked up a different soup of subatomic particles and atoms. (Mirror matter would have affected the expansion rate of the universe, in turn modifying the cooking temperature.)

But perhaps, Teplitz said, a mirror universe might still exist—if the "mirror" is sufficiently distorted, sort of like in a carnival funhouse. Suppose that mirror particles are a little bit heavier than the ordinary particles that physicists are used to playing with. A mirror electron might be 10 to 100 times heavier than the garden variety that people use to send e-mail. Protons and neutrons might be more massive than their nonmirror cousins as well.[26]

Such an overweight neutron would be unstable, decaying away even if trapped in an atomic nucleus. Thus this version of the mirror world could contain no atoms other than hydrogen, ruling out mirror people. But space could still contain interesting mirror-matter structures, perhaps helping to solve some of the deepest mysteries in the cosmos.

In their March 1996 paper, Teplitz and Mohapatra identified three possibilities: One, globs of mirror matter might have formed in the early universe and are now large and puffy; two, the puffy globs could have cooled rapidly into "cluster" globs containing large chunks of mirror matter; or three, puffy globs cooled slowly to form "black globs," condensed regions of mirror matter mimicking massive black holes.

It is feasible, Teplitz and Mohapatra calculated, that a million mirror-matter globs originally occupied the Milky Way galaxy and that some of them (say 10,000 or so) escaped into intergalactic space. "Thus one might consider searches for . . . globs both inside the galaxy and outside," they wrote.[27]

MIRROR MACHOS

But how to search for the invisible? One way would be finding distant stars traveling along paths that seem to be deflected by an invisible glob-sized mass. (Such deviations in the path of the planet Uranus were the clues used by astronomers to discover the planet Neptune.) A similar strategy has been used by astronomers to search for those MACHOs wandering about the outer edges of the Milky Way galaxy.[28] In their paper, Teplitz and Mohapatra implied that collisions of mirror-matter globs could disperse some bodies into space that might masquerade as MACHOs. But they didn't pursue the idea at the time (although a paper by Berezhiani discussed the mirror MACHO possibility in a little more detail).

In December 1998, though, Teplitz heard a presentation in Paris by astrophysicist Katherine Freese of the University of Michigan. Her analysis basically ruled out all the ordinary MACHO possibilities. Teplitz then applied Sherlock Holmesian reasoning to the problem—if you eliminate everything else, whatever remains must be the answer, even if it's something as crazy as mirror matter.

So he and Mohapatra began to calculate away. Assuming a mirror mass distortion factor of 15—in other words, mirror particles with masses 15 times that of their ordinary matter counterparts—the physicists investigated the formation of massive mirror-matter stars and their subsequent explosion, which would leave mirror black holes behind. (Ordinary black holes form when stars much more massive than the sun explode and then collapse under the force of their own gravity.) Because of the mirror mass distortion factor, a mirror black hole would not be heavier than the sun, but lighter, maybe half a sun's mass or so—just the right mass for MACHOs, the calculations showed.[29]

"This is for the theorist the same thrill that the biologist gets when the new chemical kills all the germs in the whole damn petri dish, and a lot in the next one over to boot," Teplitz told me.

This result didn't exactly prove anything, of course, but it made the possibility of mirror MACHOs more interesting.

More recently, others have proposed even more interesting mirror-matter possibilities. Robert Foot and Ray Volkas, of the University of Melbourne in Australia, have pursued a mirror-matter agenda that differs in key respects from that of Teplitz and Mohapatra. In the Australian version, the mirror particles possess precisely the same mass as their normal counterparts. This idea has the advantage of restoring the original mirror-matter motivation of overall mirror symmetry but assumes some way can be found to elude the problems posed by the element-cooking temperature considerations in the early universe.

In any event, Foot and Volkas don't stop with mirror MACHOs. They see signs of mirror matter almost everywhere. Foot has gone so far as to suggest that mirror matter has unwittingly been detected already, in the form of mirror planets orbiting distant stars.

By the beginning of 2002, astronomers had spotted telltale signs of more than 70 planets around stars far from the sun. Nobody can see these planets, of course—they're much too far away for even the Hubble telescope. But the light from a distant star is distorted by the presence of a planet. As it orbits, the planet tugs slightly on its star (because of gravity), pulling the star a little bit away and then a little bit toward Earth. The to and fro motion alternately changes the colors of light reaching Earth-based telescopes. If a planet really is the cause, the color change would follow a regular schedule corresponding to the planet's orbit; from the regularities in the pattern, astronomers can deduce something about the size of the accompanying planet and its distance from its parent star.

In one paper, Foot argued that nobody could know for sure what those planets are made of. Perhaps even if you hopped into the Starship Enterprise and cruised to the vicinity of such a star, you'd see nothing in orbit at all—if the planet were made of mirror matter.

A similar possibility arose in October 2000, when astronomers reported 18 large planets wandering through space in the constellation Orion without any stars around. It was hard to explain—after all, astronomers believe that planets form in the debris surrounding a star following its birth. No star, no debris. And therefore, presumably, no planets.

"This new kind of isolated giant planet . . . offers a challenge to our understanding of the formation processes of planetary mass objects," astronomer Maria Rosa Zapatero Osorio and collaborators reported in the journal *Science*.[30]

One possible explanation was that the "planets" weren't planets at all, but fizzled stars known as brown dwarfs. Brown dwarfs are much bigger than planets but not quite big enough to generate the internal pressure needed to burst into starhood. Zapatero Osorio, of the Instituto de Astrofísica de Canarias in Tenerife, Spain, and her colleagues estimated the Orion planets to be 5 to 15 times the mass of Jupiter, and the upper end of that range reaches the lower range for the masses of brown dwarfs. Since it was hard to deduce the mass of the Orion planets more precisely, it's possible that they belonged in the brown dwarf category.

On the other hand, there's another possibility. Maybe those orphan planets *are* orbiting stars, but the stars themselves are invisible—because they're made of mirror matter.

Foot, Volkas, and another Melbourne physicist, Alexandre Ignatiev, produced a paper proposing that the Orion planets orbit mirror stars. "Because ordinary matter is known to clump into compact objects such as stars and planets, mirror matter will also form compact mirror stars and mirror planets," the Melbourne researchers wrote.[31]

It's perfectly possible, they suggested, that ordinary matter might coalesce into a planet around a mirror star. And that would explain why the Orion planets appear to be floating freely. If the Orion planets really orbit mirror stars, the Australians said, it should be possible to detect shifts in the radiation emitted by the planets caused by the gravity of the mirror stars. If such signals are detected, the case for mirror matter as a spectacular example of prediscovery might grow just a bit stronger. Then again, somebody else might come along with a different explanation.

Still, the mirror-matter scenario offers an attractive feature that many physicists find compelling—it is built on the notion of symmetry. And symmetry, physicists agree, is super.

SUPER MATTER

3

From Noether's Symmetry Theorem to Superparticles

Without regularities embodied in the laws of physics we would be unable to make sense of physical events; without regularities in the laws of nature we would be unable to discover the laws themselves.

—David Gross
Physics Today

Emmy Noether is hardly a household name.

She'll probably never be an answer on *Who Wants to Be a Millionaire?* or a question on *Jeopardy*. She rarely comes up in conversation, even conversations where people talk about Einstein or Feynman or Marie Curie. Yet Emmy Noether was one of the great mathematicians of the twentieth century, regarded at the time of her death in 1935 as perhaps the greatest woman mathematician in history.

During her career in Germany and the last one and a half years of her life at Bryn Mawr College in Pennsylvania, she made major contributions to various fields of mathematics, particularly in advanced forms of algebra. She also showed that some of physics' most sacred laws are not accidents of nature, but rather are strict requirements—imposed by fundamental symmetries in space and time.

Take the law of conservation of energy, for example. It was discovered in the middle of the nineteenth century after a considerable amount of toil and trouble. No theorist prediscovered it. It seemed to be simply a lesson taught by observation—energy could be neither created nor destroyed. However much energy you started with, you ended with. Energy, in physics lingo, is "conserved."

In 1918, though, Emmy Noether showed that conservation of energy, and other important conservation laws, could have been deduced from purely mathematical considerations—assuming that moving through space and time did not change the laws of nature. In other words, she proved that the universe has something deep in common with snowflakes.

Snowflakes are exquisite examples of symmetry. You don't need to be a scientist to see it. Each snowflake exhibits six elaborately designed yet identical arms. Turn the flake by 60 degrees once, then again, then again. From each angle the appearance of the snowflake remains the same. And that is the essence of symmetry—change without change. A circle is symmetric because it looks the same upside down or flipped over. A baseball park is symmetric if the distances to the left-field wall are equal to the corresponding distances to right field; switching left with right leaves the distances to the fences the same. A symmetric face looks the same when viewed in a mirror.

To the artist, the architect, or the biologist, symmetry shines through the messy aspects of reality to illuminate an underlying beauty. Mathematicians regard symmetry with similar awe. And to physicists, symmetry is at the very heart of using mathematics to understand nature.

"In modern physics," write Leon Lederman and fellow Fermilab physicist Chris Hill, "symmetry may be the most crucial concept of all. . . . All of the fundamental forces in nature are unified under one elegant symmetry principle. . . . Symmetry controls physics in a most profound way, and this was the ultimate lesson of the twentieth century."[1]

Indeed, symmetry is often the critical consideration in instances of prediscovery. Dirac's prediscovery of antimatter relied on the symmetry between positive and negative energy. The possibility of mirror matter hinges on nature's respecting, at some level, the symmetry between left and right. Symmetry's success at revealing nature's secrets in the past has led many physicists to believe that it will also map the way to the future.

Specifically, many scientists foresee that the future will bring proof of a special type of symmetry that they consider to possess uncommon beauty. They call it SUSY, for the Greek goddess of beautiful symmetry. (Just kidding. SUSY stands for supersymmetry. The beauty is not in a marble sculpture but in mathematical equations.)

SEARCHING FOR SUSY

Supersymmetry comprises a mathematical framework that may spell out the secrets of nature's particles and forces. If SUSY proves true, the universe could be full of a strange form of matter so far never encountered. Exploiting the possibilities latent in SUSY, physicists have identified numerous potential prediscoveries that might resolve the dark matter mystery while solving other problems as well (or perhaps creating new ones).

So far, the evidence for SUSY is slim. The situation is much as it was in June 1999, when I encountered Neal Lane, President Clinton's science adviser, during a supersymmetry conference at Fermilab. I had just heard a presentation by Jianming Qian on the latest experimental search for SUSY at Fermilab, and Lane asked me what his conclusions had been.

"There is no experimental evidence for supersymmetry," I told Lane. And he scowled. "However," I added, "he also said there is no experimental evidence against supersymmetry, either." And Lane smiled. "That means we need to do more research," he said.[2]

A few hints have been interpreted as signs of SUSY's existence. In the early 1990s, experiments at CERN (the acronym for the European Organization for Nuclear Research, located outside of Geneva, Switzerland) indicated that the strengths of the various fundamental forces, extrapolated to what they would be at very high energies, did not seem to meet at a common point where expected. Corrections for the existence of SUSY would explain the discrepancy. Early in 2001, scientists at Brookhaven National Laboratory reported new measurements on the behavior of muons in magnetic fields. Those results also suggested support for SUSY, although later analyses called that conclusion into question.

In the absence of strong evidence either way, physicists' faith in SUSY may seem somewhat surprising. But the time and money that have been poured into SUSY searches simply reflect the incredible successes achieved in the twentieth century by following the path of symmetry. Two people stand out among the pioneers of that path—the mathematician Noether, and one of the century's premier physicists, Eugene Wigner.

EMMY NOETHER

Emmy Noether came first, but she's by far less well known than Wigner. In part that's because most of her career was devoted to pure mathematics, with little application to physics. Yet no doubt part of her obscurity reflects the difficulty women had pursuing academic careers in those days. She was born in 1882 in Erlangen, a small town in Bavaria. Her father taught math at the university there, but women were not allowed to enroll. It was possible to audit classes, with the assent of the professor, and Emmy obtained permission from some of her father's friends.

She had planned to be a teacher of foreign languages—French and English—but after auditing some math classes she changed her mind. And then Erlangen changed its policies, permitting women to earn degrees, so in 1904 Emmy enrolled as a math student and graduated with honors in 1907.

From then until 1915 she worked at the university without pay, often filling in for her father as failing health impaired his ability to lecture. During that time Emmy met David Hilbert, considered by many to be the outstanding mathematician of his day. He asked her to come to the university at Göttingen to serve as his assistant.

Soon it was clear that Emmy deserved a faculty position, and she had the support of the math department. But faculty members from other disciplines objected. If you put her on the faculty, they argued, she might then someday become a professor and therefore a member of the university senate, where women were not allowed.

Hilbert was annoyed. "I do not see that the sex of the candidate is an argument against her admission," he declared. "After all, the senate is not a public bathhouse."[3]

Noether was initially denied faculty status, but through a compromise she was allowed to lecture, in courses offered under Hilbert's name. And without pay, of course. Only after World War I did the German authorities loosen up enough to allow Noether to lecture officially. In any event, Noether's presence at Göttingen was a great help to Hilbert. In particular, he called on her to work on a problem he had encountered with Einstein's theory of general relativity, the theory that explained gravity.

Noether had arrived at Göttingen shortly before Einstein visited in the summer of 1915 to deliver a series of lectures on his new theory. (It wasn't quite finished at the time; not until November did Einstein add the final touches and figure out the proper form of the key equations.) So Noether was sufficiently familiar with Einstein's theory that Hilbert sought her input on a tricky question involving the conservation of energy.

Einstein's 1905 theory of special relativity posed no problem for energy conservation. In fact, it was easy to show that if you monitored any specific volume of space, the amount of energy flowing outward across that volume's boundary would exactly equal the loss of energy inside the volume. To physicists, that fact said that energy was "conserved locally."

But in Einstein's general theory of relativity, which incorporated gravity, the proof of local energy conservation no longer worked. This deeply concerned Hilbert. Violating energy conservation was considered a pretty serious crime. So he asked Noether to investigate the mathematics of general relativity to try to figure out what was going on.

Noether succeeded. She showed that while energy was not conserved locally, it was conserved globally—in other words, if you considered a big enough region of space, everything was fine. Energy conservation held. It was just that in smaller regions, looked at from different points of view, the measurement of energy content could differ depending on that point of view.[4]

Noether's solution came with a bonus. In working out the math she found that the key to energy conservation was an important symmetry in nature. And in fact, she found, any conservation law owes its power to a symmetry principle.[5] Thus she delivered to the physics world a deep insight into what symmetry really means. Many laws of nature are not merely arbitrary conditions imposed on how things must work, but reflections of profound properties of the universe captured in the symmetries of space and time.

One such symmetry ensures that funny things don't happen merely by a change of direction, a fact expressed by the law of conservation of angular momentum. Angular momentum is basically a measure of the quantity of spin, based on how much mass is spinning, how rapidly, over what distance. The textbook example of angular momentum conservation is the spinning ice skater. By pulling both arms in, the skater brings some mass closer to the center of the spinning. As the distance from the center is reduced, the skater's spinning speed must increase to keep the quantity of spin the same.

Besides promoting higher scores in figure skating, this law figures prominently in everything from the properties of subatomic particles to the behavior of pulsars in outer space. And it's all a consequence of spatial symmetry with respect to direction—in other words, space doesn't care which way you point.

Think back to the days before laser pointers and imagine one of those long sticks that teachers used to use to point at the blackboard. (Let's make it a wooden one, not the collapsible metallic kind.) You can be pretty sure that no matter which direction in space the teacher aimed the pointer, the stick stayed the same length. To the pointer, or anything else, it doesn't matter what direction in space you're pointing. Space is the same in all directions. The technical way to say it is that space is symmetric with respect to rotation. Noether showed that rotation symmetry guarantees that the law of conservation of angular momentum will hold true.

Noether also proved that ordinary (linear) momentum is also conserved by virtue of another symmetry of space, symmetry with respect to displacement—that is, movement from one point in space to another. In other words, any one point in space is just the same as any other point. It doesn't matter where on Earth, or in the universe, you do your experiment; the laws of nature will look the same.

In a similar way, if time is symmetric—one point in time is intrinsically no different from another—then energy must be conserved. So not only does it make no difference where you do your experiment, it makes no difference when you do your experiment. Thus, decades after experimenters discovered the law of conservation of energy, Emmy Noether showed that the experiments wouldn't have been necessary if those men had known more math. Conservation of energy wasn't prediscovered, but it could have been.

WIGNER

Physicists may have been behind the mathematicians in appreciating symmetry, but soon learned to take advantage of what the math revealed. One of the first to realize the importance of exploiting symmetry for physics was the Hungarian genius Eugene Wigner.

Born in Budapest in 1902, Wigner went to secondary school there with the slightly younger John von Neumann, who was destined to become one of the twentieth century's great mathematicians. Wigner also enjoyed math—and physics—but his father insisted on a practical education, so Eugene attended a technical school in Berlin to learn chemical engineering. While in Berlin, though, Wigner found time to sit in on many physics seminars at the university. In 1925, he went home to Budapest to work in his father's leather factory. But soon the offer of a physics research job came from Berlin, so Wigner seized the opportunity to become a scientist.

Back in Berlin, Wigner threw himself into understanding the mathematics of symmetry. On the advice of von Neumann, he mastered what mathematicians call group theory—the math on which much of modern physics has been built.

GROUPS AND SYMMETRY

Group theory is the sort of topic that makes me stop reading physics books. It seems so abstract, so obscure, and so complex that it always seemed to me impossible to simplify. Ultimately, though, I decided it was unavoidable. And guess what—it turns out not to be so bad after all. In fact, the basics of group theory are pretty simple. You just need to know a very short list of rules of what makes a group. It's not much more elaborate than having a vocalist, drummer, and guitarist.

First of all, you need to know that a group is just a set of things, which is pretty much in line with its common definition. In a mathematical group, the "things" might be objects, or numbers, or operations—like rotations. The key feature of a group is that its members are governed by rules that relate the members to one another in specific ways. Here they are:

Rule 1: You can combine two members of the group to produce another member of the group. (Example: 2 and 3 are members of the group; they can be combined by a procedure called multiplication that yields 6, and 6 is also a member of the group.)

Rule 2: When combining three members, you can combine the first two and then the third, or combine the second two first. (In other words, combining 2, 3, and 5 by multiplication gives the same answer if you first multiply 2 times 3—to get 6—and then multiply by 5, or if you first multiply 3 times 5—to get 15—and then multiply by 2. The answer is 30 either way.)

Rule 3: You can do something that changes nothing. (In multiplication, you can multiply any member of the group by 1, and the answer is the same member you started with. This is called the identity rule.)

Rule 4: You can undo whatever you've done. (This is called the inverse rule. You can undo the multiplication of 3 by 5 (15) if you multiply again by the inverse of 5, 1 over 5—or one-fifth. One-fifth of 15 is 3, the original member.)

Remember, groups can involve things other than numbers. Operations such as rotating geometrical figures work the same way, and in such cases the groups are referred to as symmetry groups.

Symmetry groups can get pretty complicated mathematically,

but the basic idea is the same as the symmetry of snowflakes. Remember, you can rotate a snowflake by 60 degrees and it looks the same. Rotating it by 120 degrees also leaves it looking the same. So does rotating it by 180 degrees. So you can see that these rotations make up part of a group—a combination of 60 and 120 degree rotations, both members of the symmetry group, produce a 180 degree rotation, also a member of the group (Rule 1). And you can easily check to see that you could combine the rotations in different ways to satisfy Rule 2.

You can also rotate by 360 degrees, which not only leaves the snowflake looking the same, but also returns all the arms to the original positions. In other words, rotating by 360 degrees is the same as doing nothing (Rule 3). Finally, you can undo what you've done just by rotating a negative number of degrees (counterclockwise instead of clockwise) to satisfy rule 4.

Of course, not everything in nature is a snowflake. Different objects possess different kinds of symmetries, and therefore manipulations of those objects are governed by different symmetry groups.

Wigner found that a major key to making progress in physics was figuring out which symmetry groups describe nature. He was able to show how the properties of matter's basic particles could be related to certain sets of symmetry operations. Symmetry groups, he determined, captured patterns in the laws of nature that described how elementary particles and forces interact. Instead of symmetries of rotations, elementary particles obeyed symmetries of interactions. Those interactions are governed by forces that can be described mathematically by symmetry groups.

Wigner's work made it clear that symmetry groups captured something profound about the construction of nature. The laws of nature are useful because they express regularities in the events and processes in the universe—processes that seem irregular and complex because the laws act on diverse initial conditions. And then when you look at all the laws, you see regularities in them, too. While laws summarized the regularities in natural processes, Wigner emphasized, symmetry principles summarized the regularities within the laws.

With this insight, Wigner anticipated the principles that laid the foundation for progress in understanding particles and forces. Over the decades that progress produced the Standard Model of particle physics, the symmetry-based equations that describe all the matter particles and force-transmitting particles in nature.

By the 1970s, the essential components of the Standard Model were in place. It grouped all matter into two main types of particles:

The Standard Model
(electric charge in parentheses)

Matter Particles (Fermions)

Quarks	Leptons
up (+2/3)	electron (–1)
down (–1/3)	electron neutrino (0)
charm (+2/3)	muon (–1)
strange (–1/3)	muon neutrino (0)
top (+2/3)	tau (–1)
bottom (–1/3)	tau neutrino (0)

Force Particles (Bosons)

Electroweak Force	Strong Force
photon (0)	gluon (0)
W^+ (+1)	
W^- (–1)	
Z^0 (0)	

---

quarks, which make up the particles of the atomic nucleus; and leptons, which include electrons and their subatomic cousins. Forces are
transmitted by particles called bosons.

GAUGE SYMMETRY

As it turns out, the Standard Model is based on a peculiar type of
symmetry, known as gauge symmetry. Gauge symmetry is even harder
to explain than groups. It has to do with how to reconcile different
ways of measuring (gauging) nature. Fortunately, you don't need to
learn the intricacies of gauge symmetry to get the basic idea. Remember, symmetry means that when something is changed, something
else remains the same. In gauge symmetries, what changes is the
gauge, or the system of measurement.

Gauge symmetry is a little like exchanging money when you va-
cation in Europe; you have to change the way of measuring money
but presumably you get equal value in the exchange (corresponding
to the laws of nature staying the same when you convert from feet
and inches to meters and centimeters.) But it turns out that preserving gauge symmetries requires a mechanism for managing the con-
version—something needs to tell the bankers what the exchange rate
is. Nature's way of doing this is what's commonly called "force." The
forces in the Standard Model are nature's way of enforcing gauge
symmetry—so the laws stay the same no matter what you are doing
or what units of measurement you are using.[6]

Einstein's general theory of relativity, it turns out, is a gauge
theory, and the force it requires is gravity. Its essential feature is the
ability to describe the laws of nature in the same way for any observer in motion. Another way of saying it is that the laws must look
the same no matter what kind of a map you set up to specify the
location of moving objects and observers. Such maps are known as
coordinate systems, kind of like the system of latitude and longitude

used for locating positions on the surface of Earth. In general relativity, the coordinate system describes locations throughout all of space and time.

There is no reason, of course, why any two observers should use the same system of coordinates. My sister might want to use a coordinate system centered on Avon, Ohio. An astronomer might choose the sun. Some gaseous-cloud life form orbiting Proxima Centauri might prefer to use the center of the Milky Way galaxy. If these beings wanted to communicate, they would have to transform measurements from one coordinate system to another. General relativity guarantees that such a change in gauge leaves the laws of nature the same—that is, it encompasses a gauge symmetry.

As Edward Witten has explained to me, this feature of general relativity essentially answers the question of why gravity exists. For the laws of nature to remain the same no matter how you're moving and what coordinate system you adopt, some force must be at work to convey the connection between one viewpoint and another. In general relativity, gravity is that force. Other gauge-symmetric forces govern the fundamental particles of nature. For the laws describing the fundamental particles to remain the same for everybody, forces must exist.

The first major work to apply gauge principles to particles was a historic paper published in 1954 by Robert Mills and Chen Ning Yang (the same Yang who collaborated with Lee on parity violation two years later). Following the gauge trail blazed by Yang and Mills, physicists produced the Standard Model of particle physics by the mid-1970s. By far the biggest breakthrough during that time was the use of symmetry principles to unify the math describing electromagnetism and the weak nuclear force (responsible for some forms of radioactivity)—and in the process prediscovering some unknown subatomic particles.

During the 1960s, symmetry principles were theorists' chief guide to the proliferation of subatomic particles discovered in the 1950s. "If you knew that the laws of nature looked the same from different points of view, you could make predictions that could be tested," Steven Weinberg, of the University of Texas, remarked in a 1997 interview. "Even if you didn't understand the forces involved, you didn't know where the particles came from, you could make predictions. And sometimes they would be right."

When the predictions went wrong, theorists could try out new symmetry principles to see which ones nature obeyed. "We learned a lot about what symmetry principles governed the laws of nature," Weinberg recalled.

But symmetry was not simple enough to reveal all the answers instantly. Many of nature's symmetries were not quite exact. It seemed that nature liked perfect symmetry in principle, but imperfect symmetry in practice. In the language of the physicists, symmetries were broken.

"A great breakthrough was the idea of broken symmetry," Weinberg told me. The concept originated in studies of solid-state physics, describing such phenomena as magnetism and superconductivity. The underlying idea is simple enough—something happens in the course of events to mask the underlying sameness that symmetry preserves. Remember, your face in a mirror looks pretty much like you look to other people. But if you part your hair on one side or the other, the images become distinguishable—the symmetry is broken. Something similar happens to a perfectly symmetric cloud of steam as it cools. Water droplets begin to form and then sooner or later you'll also get some ice—three forms of the same substance, breaking the symmetry of the original steam cloud.

It's the same with a magnet. Heat a magnet up, and at some temperature it will lose its magnetism. In other words, at high temperatures the magnet possesses a symmetry; particles within it are oriented in no special direction. But cool the magnet down again and the particles will line up along an axis, pointing through the magnet's poles. The magnet now singles out one direction as special—it's no longer perfectly symmetric. The symmetry has been broken.

Weinberg, in his Nobel Prize lecture, said he "fell in love" with the idea of broken symmetry. He soon figured out how to apply it to the problem of subatomic particles. "The idea was that you could have a physical system that is governed by laws that have a high degree of symmetry, and yet the symmetry won't be apparent in the phenomena, the particles," Weinberg explained to me. "To put it a little bit more mathematically, the equations have a symmetry, the solutions of the equations don't have that symmetry."[7]

In 1967, Weinberg saw that the weak nuclear force could be described mathematically the same way as electromagnetism. In other words, an underlying symmetry united the two forces, a symmetry that is broken under current conditions in the universe.

Working out the math, Weinberg found solutions corresponding to four force-carrying particles. One was massless and seemed obviously to be the photon, the particle that transmits electromagnetic force. But the other three particles were unknown at the time. Those particles turned out to be the carrier particles for the weak nuclear force. One should have a negative charge, one a positive charge, and one no charge at all. The charged ones were called W bosons and the neutral one became known as the Z boson, or Z-zero. Exactly those three particles were discovered at CERN in 1983, once again establishing the power of mathematics to produce prediscoveries. (Weinberg had already won the Nobel Prize by then, thanks to indirect evidence persuading everyone that the particles had to exist. He shared the 1979 Nobel with Abdus Salam, who had published similar conclusions at about the same time as Weinberg, and Sheldon Glashow, another contributor to the physics of the Standard Model.)

So in the Standard Model, electromagnetic forces become just one form of a more fundamental "electroweak" force. The massless photon transmits electromagnetism. But the particles transmitting the weak force, the W and the Z, are very massive. At some point in the history of the universe, the W and Z and photon all weighed the same, but then that symmetry was broken. Even earlier, scientists surmise, all the forces were equal in strength, and the particles transmitting them were all equal in mass. But as the universe cooled, the symmetries were broken to produce the four different-strength forces in the universe today, much in the way cooling steam produces three different versions of H_2O.

By the mid-1970s the Standard Model had been pretty much pieced together, with the strong nuclear force joining the electroweak. Then came the job of testing the model, a process requiring another 20 years or so. By the end of the twentieth century, the experimental evidence favoring the Standard Model was overwhelming. Frank Wilczek, a physicist at MIT, proclaimed that the name should be changed. Henceforth, he proclaimed, the "Standard Model" should be known as the "Theory of Matter."[8]

To be sure, one piece of the puzzle remained missing. The symmetries of the Standard Model could explain the existence of many subatomic particles, but not why they had mass. During the 1960s, several physicists noticed that some unknown field permeating space might solve that problem. Particles interacting with that field would seem to acquire mass, in much the way a marble trying to pass through molasses seems to acquire additional inertia. Various species of particle would acquire different masses depending on how strongly they interacted with this invisible field.

One of the physicists who figured this out, Peter Higgs of the University of Edinburgh, realized that if such a field existed, you should be able to make particles out of it. Such a particle, now known as the Higgs boson, is widely (if tritely) regarded by many as the Holy Grail of modern physics. It is a potential prediscovery that most physicists fervently, almost desperately, believe will happen. Near the end of 2000, experimenters at CERN reported a hint of the Higgs—just before the particle accelerator there was shut down to make way for a new, more powerful accelerator. So the race for the Higgs is now under a yellow flag, possibly providing an opportunity for SUSY to be found first.

SUPERSYMMETRY

At the same time the theory of matter was being developed, another approach rooted in notions of symmetry had been following a parallel path. That path's destination, many physicists hoped, would be the supersymmetric world beyond the Standard Model, the place with answers to the questions that the Standard Model couldn't answer.

Edward Witten, one of the world's top SUSY experts, explains supersymmetry as the quantum version of Einstein's relativity. "I've often thought about how supersymmetry can be explained to the public," Witten told me during one of my visits to Princeton. "Maybe there would be more enthusiasm from the public for particle physics if we could make supersymmetry sound as exciting as it is. Supersymmetry is really the modern version of relativity."

Einstein's theories of relativity seized on the realization that moving through space also means moving through time, and the secret to finding the underlying symmetry is considering space and time combined into "spacetime." It was the mathematician Hermann Minkowski who showed, soon after Einstein's original relativity papers were published, that the theory revealed important symmetries in time and space. A few years later Einstein himself showed how the special theory, limited to uniform motion, could be "generalized" to incorporate accelerated motion. And since falling in a gravitational field is, in fact, accelerated motion, Einstein's general theory of relativity was able to explain gravity as the result of the way that matter distorted spacetime.

But relativity is a classical theory. It doesn't include (and has resisted the incorporation of) the quantum features of reality that rule the realm of the atom. Nowadays, physicists realize that spacetime must have its quantum aspects, and supersymmetry may explain them. As Witten put it, "Supersymmetry is the beginning of the quantum story of spacetime. . . . It's a new symmetry involving new dimensions where you can't explain either the dimensions or the symmetry unless you know about quantum mechanics."[9]

Supersymmetry's new dimensions are utterly unlike the ordinary dimensions of space that you can move around in. It's not even that SUSY's dimensions are just very small so that you could only move around in them if you were a subatomic-sized flea. SUSY's "quantum dimensions" are smaller than small—they have no size at all. You couldn't move around in them no matter how small you were.

But SUSY's strange new dimensions bring with them one tangible physical effect—a new subatomic partner for every kind of particle now known. Because "supersymmetric partner particle" is a mouthful, most physicists call them superpartners. Or sparticles. I like to call them supermatter.

SUPER MATTER

In Einstein's relativity, the world is still the same when you interchange space and time. In supersymmetry, the world is still the same when you interchange matter with force. It's this deep symmetry between matter and force that gives SUSY the power to create new particles beyond those found in the Standard Model.

Basically, the Standard Model describes two kinds of fundamental particles—roughly, particles of matter and particles that transmit forces. An electron is a matter particle; a photon is a force particle, responsible for electromagnetic interactions. Fundamental matter particles are called fermions; fundamental force particles are called bosons. The defining feature of a boson or fermion is its spin; some composite matter particles are actually bosons. To the ordinary (bosonic) dimensions of space and time, SUSY adds "fermionic," or what Witten likes to call "quantum," dimensions.

SUSY's assertion of force-matter symmetry suggests that in some way, force and matter are just two aspects of the same thing. If so, then it ought to be possible to devise a mathematical framework describing a partner force particle for every matter particle, and vice versa. It's pretty much the same reasoning that gives every particle an antiparticle and every particle a mirror partner. And that's just what the pioneers of SUSY did. They worked out the math for a universe containing supersymmetric partner particles for every matter and force particle.

It's interesting that the early investigators had begun to develop SUSY math even before physicists had put the pieces of the Standard Model together. SUSY was born around 1970, a few years before the Standard Model really took shape. The first SUSY steps came in Russia (in those days, the Soviet Union). Evgeny Likhtman, working with Yuri Golfand at the Lebedev Physical Institute, produced the first mathematical expression for force-matter symmetry and speculated whether the equations might correspond to new particles in nature. At about the same time, Pierre Ramond (now at the University of Florida) uncovered some mathematical insights creating a stream of thought that merged with later SUSY theories.[11]

Then came an important paper in 1973 from Julius Wess and Bruno Zumino, in which the idea of superpartner particles first clearly appeared. The term *supersymmetry* itself apparently first showed up in a 1974 paper by Abdus Salam and John Strathdee.[10] But the full implications appeared in sharper detail in 1981, when Savas Dimopoulos and Howard Georgi produced a paper laying out the SUSY version of the Standard Model, with the whole shebang of superparticles. SUSY had been unveiled, with the shocking implication that perhaps physicists had been playing around with only half the particles that nature possessed.

Dimopoulos, now at Stanford, is one of the most exuberant of theoretical physicists, a fast and animated talker, clearly passionate about every sentence he utters. It's not hard to get him going. At dinner one evening, during a conference where he had been presenting newer work, I asked him about the original proposal of the SUSY world.

"People ask me how did you dare propose the supersymmetric standard model when it doubled the number of particles in the universe? It predicted particles for which we have no evidence," he said. "But I didn't find it that revolutionary."

The reason, he explained, was his familiarity with the history of prediscovery. "I knew that twice in history this had happened before," he said. "First with Dirac, predicting antimatter. Then with Pauli, predicting spin."[12] Indeed, Dirac's prediscovery of antimatter was no less ambitious than that of supersymmetry—for every known particle there would be an antiparticle, another case of doubling the census count in the subatomic universe. Pauli's accomplishment was similar, if not quite as dramatic. Electrons had previously been considered all identical; Pauli identified a distinction—some spin in one direction, others spin the opposite direction. In a sense, they could be thought of as different particles, too.

There was one big difference, Dimopoulos acknowledged. The symmetries exploited by Pauli and Dirac were exact. The mass of an antiparticle, for example, would be precisely the same as that of the ordinary particle. But superpartners could not be identical in mass. They had to be much more massive; otherwise, they would have been discovered already.

Thus Dimopoulos and Georgi had to propose that SUSY was not

perfect after all, but—at least as it appeared in nature—had to be a "broken" symmetry, like the symmetry describing the electroweak force discovered by Weinberg and Salam. Applying the idea of symmetry breaking to SUSY explained why the supermatter partners had not yet been discovered. They must be much more massive than their partners, and so it would take very high energy to produce them, beyond the power of the best atom smashers available.

DESPERATELY SEEKING SUSY

For the last two decades, mathematicians and physicists have spent countless hours developing variations of SUSY mathematics, seeking insights that would lead to explanations for known phenomena and solutions to subatomic and cosmological problems. In fact, working out the intricacies of SUSY math seems to occupy every waking moment of dozens of physicists around the world. Out of all that effort come numerous surmises about things the world might possess if SUSY turns out to be true. In other words, SUSY is a fertile field for cultivating prediscoveries.

The most likely SUSY discovery, of course, would be one of the superpartner particles. SUSY scientists are well prepared for this discovery, as names for the new particles have already been devised. For matter particles, the naming rules are simple—put an s in front of the name. Thus the superpartner of the electron would be called a *selectron;* quarks' partners would be *squarks*. For force particles (bosons), add -*ino* to the basic name—the photon's superpartner goes by *photino*, for example.

Naming the particles was the easy part. Finding them will be harder. Not only must the superparticles be much more massive than their ordinary counterparts, and therefore hard to make, they would also be difficult to detect. Despite their mass, supermatter particles would be very reluctant to interact with ordinary matter, entering only into reactions where the weak nuclear force is involved. These timid, *weakly interacting massive particles* are known as WIMPs.

SUSY says WIMPs of all sorts should exist—one type of WIMP for every known type of particle. Right after the big bang, WIMPs should have been abundant. But they would also almost all be unstable, decaying into lighter particles, so most of the WIMPs in the universe would be long gone by now. But one of them has to be the lightest of all, and it should still be around, in massive quantities, if SUSY is true.

Sometimes physicists call it the lightest supersymmetric particle, or LSP. Since it would certainly have no electrical charge, some theorists think it is probably the photon's superpartner, the photino. On the other hand, some physicists prefer to refer to the LSP as the *neutralino*, because it might actually be a quantum mixture of different neutral superparticles. (In quantum physics, you cannot specify a particle's identity with absolute certainty. The equations allow a given particle in flight to possess properties of different related species simultaneously. When you capture it, it adopts one identity.)

In any event, WIMPs may very well be abundant in the universe, flying freely through space, a few passing through the very room you're sitting in at the moment. Since they interact weakly, though, you are in no danger—although physicists seeking a sign of the WIMPs might be willing to risk a little danger to improve the chances of finding one.

Actually, some searchers think they have already succeeded.

You might remember from Chapter 2 (this is only Chapter 3, after all) that astronomers infer the existence of a lot of mass in the outer regions, or halos, of galaxies. Some of it seems to be in the form of *massive compact halo objects*, known as MACHOs. But not all of it. Most experts believe that at least some of the dark matter comes in the form of WIMPs.

In fact, at the Texas Symposium for Relativistic Astrophysics in

Paris in 1998, it seemed that the WIMPs were about to kick the *H* out of the MACHOs. At that meeting Katherine Freese, of the University of Michigan, suggested that the sightings of MACHOs between Earth and the Large Magellanic Cloud may have been misleading. It's possible, she said, that the MACHOs were not in the halo after all, but in the Large Magellanic Cloud itself. Another MACHO candidate, seen toward the Small Magellanic Cloud, was almost certainly in the cloud, not the Milky Way halo, she said. If so, maybe the dark matter was mostly WIMPs, and MACHO should be rewritten as MACO. Which doesn't have quite the same ring to it as MACHO.

Freese suggested that as much as 90 percent of the galactic dark matter is WIMP matter. And at the same meeting, a team from the DAMA (for *dark matter*) experiment in the underground Gran Sasso laboratory in Italy reported a strong hint of a particle that matched a WIMP's expected properties.

If WIMPs lurk throughout the galaxy, the DAMA team reasoned, the Earth should be running into them all the time. After all, the whole solar system speeds around the galaxy at 140 miles per second. And if WIMPs really are the dark matter, there ought to be maybe one WIMP particle out there in every cubic centimeter of space. With that many WIMPs, it ought to be possible to detect some of them, even if most escape notice. So the DAMA experimenters constructed detectors containing chunks of sodium iodide that give off a flash of light when a WIMP strikes.

Even though the detectors are deep underground, to screen out other kinds of particles that might fool the detectors, it's impossible to know whether any given flash really represents a WIMP encounter. There might be radioactive rocks somewhere giving off particles of some sort as well, for example. Presumably, though, any other particles that strike should do so all year round, with no preference for summer over winter. WIMPs, on the other hand, would strike more often in the summer, when the Earth is moving through the galaxy in the same direction that the sun is—into the WIMP wind, so to speak. In the winter, the Earth is revolving away from the WIMP wind. Therefore, if the detector flashes more in June than in December, the extra flashes may be signals of WIMPs.

And that's just what the DAMA team reported in Paris. In 1997, Pierluigi Belli of the DAMA team reported, the detectors saw a hint of the excess in June. In 1998, he said, the team found an even stronger signal. Belli said the team's analysis favored a WIMP weighing in at about 59 billion electron volts, or roughly 60 times the mass of a proton.

In the question period following his talk, however, other scientists sharply criticized the DAMA team's data analysis. Similar disputes arose a year later during a conference in California. The DAMA team once again proclaimed their belief that WIMPs had been detected. This time, though, a rival team disputed their analysis. Experiments at Stanford, using a different WIMP-searching method altogether, had also recorded some flashes in their detectors. But those flashes were not caused by WIMPs, the experimenters concluded, but by neutrons.

By the end of the year 2000, the controversy had not cooled. In December, the Texas Symposium on Relativistic Astrophysics was actually held in Texas for a change—in Austin, where the WIMP debate continued. Rita Bernabei, leader of the DAMA team, delivered a spirited and contentious defense of her group's findings. After four years of tests, she proclaimed, the June-December mismatch in detections remained clear.

"Where you expect the maximum you get the maximum," she said. "Where you expect the minimum you get the minimum." Possible confusion from other particles, say, radioactive emissions from underground radon, could be excluded, she said. "We have the presence of a modulation with the proper features for WIMPs."[13]

Blas Cabrera, of the coalition performing the WIMP search at Stanford, was not impressed. He remained calm, but clearly rejected Bernabei's claim. "Our conclusions," he said, "are in disagreement with those of the Rome group."[14]

Rather than comparing June and December, he explained, the Stanford experiment tried to trap WIMPs in small detectors made of silicon or germanium, maintained at ultracold temperatures. Both are semiconducting elements used in electronic devices, and both are sensitive to the impact of WIMP particles. In the case of silicon, sensors are tuned to faint vibrations caused by a WIMP impact. With germanium, the sensors measure the tiny rise in temperature caused when a WIMP deposits its energy.

The Stanford experiment was set up only 35 feet underground— shielded from most problems, but not deep enough to escape an occasional cosmic ray. Cosmic ray particles called muons could, without too much trouble, smash into rocks outside the experimental chamber and eject neutrons that would trigger the germanium or silicon sensors. There would be no obvious way to tell if any given impact had been caused by a neutron rather than a WIMP.

However, the germanium detectors are very much more sensitive to WIMPS than silicon is, while silicon is only a tiny bit more sensitive to neutrons than germanium is. Consequently, if both types of detector record hits at about the same rate, they must be seeing neutrons, not WIMPs.

And that's just what the results seemed to indicate—both detector types recorded something like one or two hits a month. If the hits were from WIMPs, germanium should have been recording ten times as many hits as silicon.

"Our conclusion is it's a better fit with a neutron background than a WIMP signal," Cabrera said. But if the June-December effect seen by DAMA really revealed the presence of WIMPs, he said, the Stanford experiment should have seen some, too. Cabrera attempted to remain diplomatic and noncommittal in his statements, but the clear implication was that the DAMA team might have committed some errors in its analysis.

Fortunately, further WIMP searches are in the works. The Stanford group plans to move more sensitive equipment to a mine in Minnesota, where greater depth below the surface will reduce contamination from cosmic rays, and any WIMP signal should emerge more clearly. Atom smashers continue to probe higher ranges of mass where WIMPs might be found. WIMPs may even turn up in collision debris at Fermilab, the Illinois atom smasher, any day now. An even better bet, most scientists think, is the Large Hadron Collider, now under construction at CERN, expected to begin smashing in 2006 or so.

If SUSY particles don't show up by then, many scientists will be disappointed, and even surprised. Physicists have long known that all the dark matter in the universe cannot be made of MACHOs. Some dark matter must be something other than the ordinary stuff from which MACHOs presumably are made.

So if WIMPs are not found, many physicists will be frightened. For that would raise the likelihood of a different potential pre-discovery: the idea that dark matter might actually be the terrifying particles known as WIMPZILLAS.

DARK MATTER

4

From Pauli and the Neutrino to the Universe's Missing Mass

The discovery by Zwicky that visible matter accounts for only a tiny fraction of all the mass in the universe may turn out to have been one of the most profound new insights produced by scientific exploration during the 20th century.

—Sidney van den Bergh
"The Early History of Dark Matter"

For cosmologist Rocky Kolb, size does matter.

To him the idea that most of the matter in the universe is made of particles called WIMPs is somehow a source of embarrassment. True, WIMPs are the most likely members of the supersymmetry side of the subatomic family tree to be found in space. They may very well account for much of the mysterious dark matter in the universe.

But maybe not. Some computer simulations suggest that a universe full of WIMPs would not produce the right number of small "dwarf" galaxies that surround big galaxies like the Milky Way. Other nagging inconsistencies suggest that the universe could not be built by WIMPs alone. As a result, the identity of the dark matter remains mysterious, prompting Kolb and others to propose a zooful of novel species of matter to populate the cosmos.

Peruse the astrophysics literature and you'll find more candidates for dark matter's identity than remakes of Godzilla films. Proposals include large bodies like black holes, brown dwarfs, red dwarfs, and white dwarfs; massive quantities of small particles, like WIMPs, axions, or strange quark nuggets; and more exotic speculations, like mirror matter or cosmic Q-balls.

And then there's Rocky Kolb's favorite candidate. He calls them WIMPZILLAS. He can get away with what seems an outlandish suggestion for a good reason—nobody else really has the slightest idea of what the dark matter really is.

It's been that way since the 1930s, when Fritz Zwicky, a cantankerous Caltech astronomer, noted some strange behavior in a group of galaxies known as the Coma Cluster. Those galaxies moved across the sky with a speed that simply couldn't be explained if they made up all the matter in the cluster. In 1933 Zwicky reported that the galactic motions implied that the Coma Cluster contained a lot of matter that astronomers couldn't see. "If this . . . is confirmed," he wrote, "we would arrive at the astonishing conclusion that dark matter is present with a much greater density than luminous matter."[1]

Later, observations of other clusters confirmed the discrepancy—more mass appeared to be present than the amount that visible galaxies could account for. A further dark-matter mystery arose in 1939, when Horace Babcock measured how fast the outer region of the Andromeda galaxy was spinning. He found that stars on Andromeda's outer edges appeared to revolve around the galaxy much more rapidly than they should, based on simple applications of Newton's law of gravity.

Farther-out stars should be revolving more slowly, just as Pluto, the farthest planet from the sun, orbits at a much more leisurely pace than the innermost planet, Mercury. It turned out that not only Andromeda, but other galaxies as well, rotated just as fast on their outer edges as they did much closer to their centers. Apparently, the mass of those galaxies increases with distance from the center, while their brightness does not. The only plausible explanations are that something is wrong with the law of gravity (which few physicists think is likely) or that the visible part of a galaxy is embedded in a vast massive halo of unseen (that is, dark) matter.

At first, nobody seemed to make the connection between this missing matter around galaxies and the missing matter in the Coma Cluster. Astronomers in general did not worry about dark matter much at all until the 1970s. But then further studies by the astronomer Vera Rubin and colleagues found more and more galaxies with high outer rotation rates. Observations of other clusters confirmed Zwicky's suspicions as well. By the 1980s it was well-established that 90 percent or so of the mass of a typical galaxy is unseen and that massive amounts of dark matter lurk both in galactic halos and in the vast spaces between galaxies as well.

In a way, it's pretty amazing that after all this time, astronomers cannot say what this dark matter is made of. It's one of the greatest mysteries in the history of science, or perhaps in the history of anything. Imagine living in a house and having no clue to what it is made of. Or realizing that the inside of your body is something other than skin, but not having any idea what. You'd want to know. Astronomers and physicists desperately want to know what the universe is made of, too. And here's a prediction: when scientists finally do find out what the dark matter is, it will be something that somebody has already predicted. A prediscovery.

No realm of physics and cosmology provides a more fertile field for prediscovery than the dark-matter mystery. In fact, one candidate for contributing to the dark matter is itself one of science's greatest prediscoveries, the ghostlike particle known as the neutrino.

PAULI AND THE NEUTRINO

In the form of radioactivity known as beta decay, an atomic nucleus shoots out an electron. (Electrons emitted in this way are therefore called beta particles.) Careful measurements show that the electrons that fly away do not always possess the same amount of energy. Even if you account for the energy of the motion of the atomic nucleus they come from, these electrons can still exhibit a range of energies. But the total amount of energy in a process is supposed to remain constant, as the law of conservation of energy requires.

When radioactivity was discovered, at the end of the nineteenth century, the law of energy conservation was only a few decades old. Some scientists suspected that perhaps that law had been repealed by radioactivity. Or maybe it was just an approximate law that was un-enforceable on the atomic scale. But plenty of experimental evidence argued otherwise. Finally an alternate solution was proposed by one of the most critical thinkers of his era, the Austrian physicist Wolfgang Pauli.

Pauli, who was born in 1900 and died in 1958, remains one of the legendary figures of physics lore. His signature scientific achievement was the Pauli exclusion principle—the limit on packing particles together that helped Dirac prediscover antimatter. Pauli was also famous for the "Pauli effect," based on his experimental ineptitude. When a lab apparatus would blow up for some unknown reason, physicists suspected that Pauli must have been passing through town at that moment. (On one occasion, while riding on a streetcar, Pauli and some colleagues witnessed a crash between two other streetcars. Pauli turned to his friends and said, "Pauli effect!")[2] Everyone was thankful that Pauli was a theorist.

More seriously, Pauli was known as the sharpest critic of new ideas among the leaders of European physics in the first half of the twentieth century. Nobody was quicker than Pauli to spot a flaw in someone else's equations. "No form of approval could be more precious to physicists . . . than Pauli's benevolent nodding," the physicist Léon Rosenfeld once wrote.[3]

Much of the time, though, Pauli's response to a presentation was not so benevolent. He was likely to blurt out something like "but this is all wrong!" Chen Ning Yang recalls delivering a seminar where Pauli interjected such virulent criticism that Yang decided to sit down and stop in mid-presentation.[4]

One joke circulated about a dream of Pauli's in which he had died and gone to heaven, where he met face-to-face with God. Pauli seized the opportunity to ask a favorite question among physicists— why a certain combination of physical quantities produced a number almost exactly equal to 137.

"Why 137?" Pauli asked.

"It's all here in these equations," God responded, handing over a sheet of paper.

Pauli looked it over for a few seconds and then said, "But this is all wrong!"

Just as surely, Pauli knew that something was wrong with the theory of beta decay. As other mysteries of quantum physics began to clear up toward the end of the 1920s, the beta particle problem became even more perplexing. Abandoning the principle of energy conservation, radical as it seemed, was actually considered a serious option by some physicists. Pauli proposed a perhaps more palatable but equally bold solution—the existence of an entirely new particle unlike anything previously known to physics. It was nothing like any ordinary bit of matter, but rather something ghostlike, a sort of stealth particle that adjusted the speed of beta particles by siphoning off the missing energy itself.

Pauli articulated his new idea in a letter sent to a scientific meeting in Tübingen in 1930. (He should have been at the meeting but elected to remain in Zurich so he could go to a dance instead.) He called the proposed particle a neutron, since it should carry no electrical charge.

"Dear Radioactive Ladies and Gentlemen," Pauli wrote in his letter, dated December 4. "I have considered . . . a way out for saving the . . . conservation of energy." Atomic nuclei must contain a neutral particle (a neutron) that would be very light, perhaps about the same mass as electrons, he explained. "The continuous beta spectrum would then be understandable," he continued, "assuming that in the beta decay together with the electron, in all cases, also a neutron is emitted, in such a way that the sum of the energy of the neutron and of the electron remains constant."[5]

Neutron turned out to be a bad choice of names, because of confusion with the particle we now call the neutron, discovered in 1932 by James Chadwick in England. But Chadwick's neutron was a massive particle, the size of a proton, nothing like the mysterious lightweight particle that Pauli had in mind.

By 1934, the Italian genius Enrico Fermi solved the nomenclature conflict in a paper working out the math behind Pauli's idea. Fermi changed the name of Pauli's neutron to *neutrino*, Italian for "little neutral one."

Shortly thereafter, the physicists Hans Bethe and Rudolf Peierls analyzed the theory to calculate how much neutrinos would interact with ordinary matter. The answer: not much. An average neutrino could zip through a wall made of lead trillions of miles thick with no problem. And that raised a serious question about how you could know whether neutrinos really existed; after all, interaction with matter of some sort would seem to be the only way a neutrino could actually be detected. Bethe and Peierls suggested that you'd have a chance of catching one in liquid hydrogen, the catch being that your liquid hydrogen tank would need to be a thousand light-years wide. (At 65 mph, it would take 10 billion years to drive that far.) Bethe and Peierls concluded that "there is no practically possible way of observing the neutrino."[6]

Pauli himself did not have very high hopes that anyone would ever prove the existence of his particle. "I have done a terrible thing," he said. "I have postulated a particle that cannot be detected."[7]

But that didn't stop Fred Reines from trying.

In the early 1950s, Reines was a young physicist at the Los Alamos laboratory in New Mexico, searching for a significant project. Finally he decided to try to detect the neutrino. Working at a weapons lab, he knew that nuclear bombs produced an enormously intense blast of neutrinos, just what you'd need to have a chance of nabbing one. After all, Bethe and Peierls had made their prediction long before such a prolific source of neutrinos was available. Reines enlisted the help of Clyde Cowan, another physicist, and they began a collaboration to show that it was possible to do the impossible.

Of course, there were some problems with setting up a detector next to an atomic bomb. Repeating the measurements would be difficult, for example. But it turned out that nuclear reactors also produce a good enough supply of neutrinos, and setting up a detector near a reactor seemed a lot easier. So Reines and Cowan altered their strategy, opting for reactors over bombs. By the mid-1950s they had succeeded, detecting the unmistakable signal of a neutrino striking a proton. After a conclusive experiment in 1956, they telegraphed Pauli that they had found the proof of his neutrino's reality.

"Everything comes to him who knows how to wait," Pauli wrote back.

Reines also knew how to wait. In 1995, almost 40 years after the experiment, he won a Nobel Prize for detecting the neutrino. (Cowan had died in 1974.)

Yet long before Reines trapped a neutrino, its existence had been taken for granted by most physicists—there simply was no other way to explain beta decay. Later it would turn out that the neutrino would have other uses—perhaps, for example, explaining dark matter.

COMPLICATING THE COSMOS

Before the 1980s, the standard view of particle physics held neutrinos to be massless. But here and there hints began to appear that maybe the neutrino had a little bit of mass after all. If so, neutrinos might make up a major portion of the dark matter in space. Even a tiny mass would add up, considering the numerous neutrinos speeding through the cosmos. (At any moment, thousands of neutrinos are zipping through your body.)

By the end of the 1980s, though, most astrophysicists concluded that neutrinos could not be the dark matter, for their speed would have worked against the need to build galaxies in the universe's youth. At most, it seemed neutrinos could possess only a tiny amount of mass and therefore would zip through space at very nearly the speed of light. In astrophysical terms, neutrinos would be "hot" particles—that is, particles that move very rapidly. (Slowpoke particles are considered "cold.") Hot dark matter did not appear to be the right ingredient to explain the galactic superstructures the universe had cooked up.

Before the 1980s, astronomers knew only that they couldn't see some of the matter out in space, and had no clue about whether the mystery matter was hot or cold. But midway through the 1980s, new observations revealed that the universe was a more complicated place than anyone had previously realized. Galaxies (or small clusters of galaxies) were not, as astronomers had generally assumed, scattered randomly through the cosmos. Instead the universe turned out to be an architectural marvel, a network of bubbles and walls stretching across all of visible space.

The bubble story popped into astronomical consciousness in 1986, when astronomers Margaret Geller, John Huchra, and Valérie de Lapparent reported their first efforts at mapping the locations of about a thousand galaxies in a slice of sky visible in the Northern Hemisphere. The astronomy world was astounded. Their map revealed a universe of richer structure than previously believed possible, showing that galaxies are not scattered throughout space but congregated along the surfaces of imaginary spheres, like giant bubbles. (The "Lawrence Welk universe," one headline writer called it.) Later Geller and Huchra found that some clusters of galaxies seem to be arranged over several bubble surfaces to form a long "Great Wall" extending 100 million light-years across the sky—bigger than the solar system to the degree that the Great Wall of China is bigger than a bacterium. Subsequently a similar structure was found in the southern sky.

Around the same time, other astronomers reported that some galaxy clusters seem to be streaming rapidly toward a massive "Great Attractor," a mysterious unseen but unusually dense region of the universe, further suggesting structure in the universe on very large scales.

The discovery of such fantastic structures presented a new challenge to astronomers trying to explain how galaxies formed in the early universe. You'd think that galaxies should be arranged at random. And you'd think so because when the universe was young, matter was spread uniformly through space, with no large lumps. Lumps from back then would show up today as cold or hot spots in the cool glow of radiation left over from the birth of the universe.

This "cosmic microwave background" radiation was generated less than half a million years after the big bang, so its features provide astronomy's equivalent of a fossil from the universe's youth. It represents a snapshot of space at an early epoch, reflecting the distribution of matter before galaxies existed. And that radiation looks very smooth: its temperature is almost exactly the same no matter what direction in the sky it comes from. Therefore the universe back then must have been filled with a smooth sea of matter, and thus galaxies should have formed at random. Just by chance, a few bits of matter would have bumped into each other to form a lump a little denser than the matter around it. A slight density advantage would then be magnified by the action of gravity, as one lump would draw more matter in. Giant galaxies would grow from those tiny matter seeds.

But galaxies did not form at random. Galaxies formed in clusters, and clusters of clusters, separated by the giant bubble-like "voids" in which relatively few galaxies are found. The seeds of matter in space must have been planted in a complicated way.

It wasn't until the 1990s, though, that astronomers detected signs of those seeds in the cosmic microwave background. Starting with the famous satellite COBE (for Cosmic Background Explorer), various measurements have shown subtle patterns of temperature differences, reflecting tiny lumps of matter from the early days. The trouble is, the lumps were too small to have grown, in the time available, into the giant structures visible today, if those lumps were made only of ordinary matter. Ordinary matter (primarily protons and neutrons) could not coagulate rapidly enough. Some other form of matter must have been present—a form that could have coagulated earlier than ordinary matter, but without disturbing the microwave radiation.

Naturally enough, astronomers suspect that the dark matter they can't see today might be the mystery matter that existed back then. If so, the dark matter cannot be ordinary (baryonic) matter.

Neutrinos are not baryons, but the dark matter can't all be neutrinos, either. Experiments showing that neutrinos have a small mass indicate that it is not enough to account for all the matter that is missing. All the neutrinos added together might weigh as much as the visible matter (basically, stars) in space, but that's only about a tenth as much as all the matter out there. Besides, neutrinos make hot dark matter. Hot dark matter would require more time to cook up large-scale clustering than the age of the universe provides.

All these developments have led many astronomers and physicists to believe in the existence of "cold dark matter," so named in the early 1980s by the cosmologist J. Richard Bond. Cold dark matter consists of slower-moving particles that seem to offer just the right

ingredient to make the galaxy clustering work out the way it is supposed to.

It turned out that WIMPs—the SUSY partner particles from
Chapter 3—would be perfect cold dark matter candidates. They'd be
heavy (otherwise they would already have been discovered), ranging
from 50 to 100 or even 1,000 times the mass of a proton. Therefore
they'd move rather slowly. And they would be weakly interacting.
That would give them just the right combination of properties to
help make the seeds that grew into galactic superclusters.

On the other hand, maybe the dark matter is something even
stranger. SUSY-WIMPs may exist, but they might not tell the whole
story. Plenty of other potential prediscoveries have been postulated
to make up some, if not all, of the dark matter. All of these hypothetical matters are pretty strange. But one of the strangest is also one of
my favorites, the curious characters known as cosmic Q-balls.

Q-BALLS

Q-balls have nothing to do with cotton swabs, billiards, or villains
on *Star Trek: The Next Generation*. Q-balls are lumps of super matter that
may have formed when tiny superparticles coagulated in the hot
dense phase of the early universe. If they really exist, it's possible
that some Q-balls lurk in the shadows of galactic halos even today,
making up at least some of the dark matter.

Q-balls offer more than a possible solution to the dark-matter
mystery. If real, they could alter the history of the universe, provide
power beyond the dreams of the Energizer bunny and produce weapons dwarfing the destructive force of mere atomic explosives. A Q-
ball-bomb could outbang the first atomic bombs as much as those
bombs outblasted a stick of dynamite.

So far as I know, Q-balls have never received any significant attention in newspapers, apart from a column I wrote about them in

1997. (The British science magazine *New Scientist* did contain an in-depth report on them in May 2000). But they do show up from time to time in talks at scientific meetings.[8] I encountered them first in several papers appearing on the Internet, such as one by Alexander Kusenko, at the time a physicist at the CERN laboratory near Geneva. When the universe was very young and hot, immediately after the big bang, Q-balls could have been produced in huge quantities, Kusenko noted in his paper.[9]

It's by no means a sure thing, but it's plausible, assuming the validity of supersymmetry. When the universe was young and hot, squarks and sleptons (the SUSY partners of quarks and leptons) could have coagulated into balls. Some such balls would have been unstable and broken apart, or they might simply have evaporated away without doing any damage. But some might have survived long enough to inject extra ingredients into the primordial soup of matter and energy. Q-balls therefore might have affected early-universe events, such as the creation of different chemical elements, in ways that standard accounts of the universe's history haven't considered.

In another paper Kusenko and a colleague calculated that some Q-balls might even survive for billions of years and are perhaps still floating through space today. "The relic Q-balls can . . . survive until present and contribute to the dark matter in the universe," wrote Kusenko and Mikhail Shaposhnikov.[10]

A lone Q-ball floating through interstellar space would be hard to see, they acknowledged. But the gravity of stars and planets might lure Q-balls to stellar or planetary centers. "It is conceivable that the deep interior of small planets might become accessible for exploration in the future and reveal storages of primordial Q-balls," the physicists said.

Well, a voyage to a small planet, or to the center of the Earth, would probably take too long to satisfy most physicists. But there are other prospects for Q-ball prospecting. More powerful atom smashers might someday be able to produce those superpartner squarks and sleptons, possibly allowing the study of Q-balls in the laboratory. If so, Q-balls might provide a new avenue for probing higher energies, as the interior of a Q-ball would provide information on energy levels far greater than an affordable atom smasher can achieve. And there might be a big bonus for society. Just as a Q-ball bomb would give 100 H-bombs' worth of bang, a Q-ball power plant would provide a practically inexhaustible supply of ordinary energy.

The best part is the Q-balls wouldn't be used up—they would just be catalysts. The fuel would be protons, abundantly available from the hydrogen in water. Shoot a beam of protons into a Q-ball, and its internal superparticles would rip each proton's quarks apart, releasing all the energy that had been holding those quarks together. You could use the energy released to boil water (the way most ordinary electric power plants do to drive steam turbines) or figure out some other scheme to tap the Q-ball energy output, Kusenko, Shaposhnikov, and Gia Dvali wrote in another Q-ball paper. "There are several processes that can yield large amounts of energy once a Q-ball is assembled and placed in a beam of protons," they wrote.[11]

Of course, the promise of cheap, inexhaustible energy has been heard before. And apart from the scientific uncertainties, the technological challenges of coping with Q-balls would argue against buying stock in any such venture at this time (although California may want to look into the possibility of Q-ball power). As the CERN scientists put it, "Technical and engineering aspects of such process, which may or may not be possible to realize in practice, lie outside the scope of our investigation."

In any event, if Q-balls exist, they would surely be one of the strangest forms of strange matter in the universe. But they would not win the title of strangest name. I award that honor to Rocky Kolb's invention, the WIMPZILLAS.

ROCKY IS NO WIMP

Rocky Kolb is my favorite cosmologist, because he's the best there is at capturing the drama and substance inside science and communicating it to people on the outside. I met him in 1991, at a physics meeting in Vancouver. Although I'd heard him talk once or twice before, it was Vancouver where I realized what a spectacular spokesman for science he was.

Let's face it, Carl Sagan is dead. Science needs people who can speak. Getting the message of science out to the public is even tougher than figuring out what the dark matter is. And very few scientists can communicate the way Rocky can. While most physicists don't know the difference between a sound bite and a snake bite, Rocky is as quotable as Will Rogers or Mark Twain.

At the Vancouver meeting, Rocky was one of three speakers in an evening session for the general public. Hundreds of visitors packed an auditorium on the University of British Columbia campus to hear what physicists were up to these days. The first two speakers treated the audience to slides showing diagrams and equations and pictures of big atom smashers. Rocky, last on the agenda, took the audience on a tour of the universe.

He explained how the young universe, just after it exploded into existence, was a dense, hot, primordial "soup" of tiny particles: protons, neutrons, leptons. "Generically, they're known in the primordial soup as croutons," Rocky said. His next slide showed a can of Campbell's soup, labeled "Primordial." On his chart showing the history of the universe, he listed important events, including the formation of atoms, formation of galaxies, birth of the solar system, and "Cubs win World Series." He showed a before-and-after slide of Supernova 1987A, a small star (indicated by an arrow on the left-hand side) that exploded into a bright spot filling most of the right-hand side.

This supernova was very important for helping scientists learn how to tell which stars will explode, Rocky said. "The stars that explode are the ones with arrows pointing toward them."

He used only one equation in the whole talk. His humor was mixed with a lot of solid science, presented in a clear way to provide people with a real sense of what science does and doesn't know about the universe and what scientists are doing to find out more.

Rocky himself is always doing something to try to find out more about the universe. He's after the big picture, the whole story of how the universe came into being and evolved into the cosmos that today's telescopes reveal to inquiring minds like his. But along the way he likes to have fun, and a big part of the fun is coming up with names like WIMPZILLAS.

It is a great name. What better way to convey the notion of a particle vastly more massive than a WIMP, a monstrous WIMP, a particle heavier than a million billion ordinary subatomic particles? And the best thing is there's an outside chance that WIMPZILLAS might really exist.

Rocky's first paper proposing the WIMPZILLA idea was hard to miss because of its catchy title: WIMPZILLAS![12] I had given it a brief mention in one of my columns, but somehow never got around to the whole story. So I was pleased to hear the way Rocky put it all in historical context during a talk at the Texas symposium in Austin in December 2000.

In a way, the WIMPZILLA story goes back to the beginning of the universe. If WIMPZILLAS are around today, they would have been created way back then, in the opening moments of the universe's existence. Their manner of birth was not imagined, however, until the 1930s, when the Austrian physicist Erwin Schrödinger put his mind to what space had been doing back at the beginning.

As most everybody now knows—although it wasn't so clear to everybody back then—space, after the beginning, was expanding.

The big bang set the universe in motion. For some reason Schrödinger was worried about this.

Back in 1939, Rocky said, you'd think that Schrödinger would have had other things to worry about. But he was concerned that the expansion of space offered a way to make matter. Thanks to quantum mechanics—which Schrödinger had played an important role in inventing—the vacuum is not a calm and quiet place on the subatomic scale. Because quantum physics allows it, particles can pop into existence out of thin space all the time. But they appear in pairs: a particle is always created along with its antimatter counterpart. That way there is no danger of so many new particles coming into existence that the universe is overwhelmed by them. Soon after their appearance, the matter and antimatter particles bump into each other and disappear in a flash, returning the energy they had borrowed from the vacuum to fuel their ephemeral existence.

Everybody seemed happy enough with this picture, but Schrödinger saw something to be concerned about. In the early universe, the universe was rapidly expanding. A particle and antiparticle, popping into existence right next to each other, might not recombine soon enough, and the expansion of space could pull them away from each other. If that happened, they would not annihilate, and the population of particles in the universe would increase by two. And in fact, Rocky says, that may very well be the process that provided the universe with the original particles it needed to make seeds for galaxies.

In truth, it's a little more complicated. "It's really the changing gravitational field that's responsible for the particle creation," Rocky explained to me. Particles get created around black holes in a similar way—as you move through space to get closer to a black hole's outer boundary, the strength of gravity changes sharply. In the first instants of the universe, the gravitational field changes rapidly in time. In either case, particles popping into existence because of quantum fluctuations are no longer able to annihilate. "If you have a rapid change in the gravitational field, like you would around a black hole, where the gravitational field changes rapidly in space, or in the early universe . . . where there's a rapid change in the expansion rate, then particle creation is more effective," Rocky said.

Specifically, the expansion rate would change very rapidly if the popular theory of inflation is right. *Inflation* is the name given to a very brief but extremely rapid puff of expansion that supposedly occurred for something like a tiny fraction of a trillionth of a second. Putting the brakes on inflation—to return the universe to a more leisurely expansion rate—required a quick and dramatic slowing of the expansion rate, a good time for producing particles.

Inflation itself requires the existence of a field, called the inflaton, to provide energy for driving the rapid expansion. A particle made from an inflaton field would be very massive. If that mass scale has some fundamental significance, then maybe other particles of similar mass should exist as well—perhaps the ones that Rocky calls WIMPZILLAS.

"Generally when you look at nature there's not just one particle of a certain mass, but it's a scale; there are many particles of that mass," Rocky said. "So if there is an inflaton that has this certain mass, if it's a fundamental mass scale, then there would be other particles of that mass. So far in our experience of nature, if you find one, you'll find another."[13]

The realization of the WIMPZILLA possibility was serendipitous, Rocky recalled. "It's something we stumbled upon," he told me when I visited him at Fermilab in May 2001.

Rocky, his student Dan Chung, and a postdoc, Tony Riotto, had been discussing ways that dark matter might have been made during the early universe. Maybe there was some connection to the inflaton, the field responsible for inflation, they decided. If some matter field interacted strongly with the inflaton, they convinced themselves, then you could show how the very massive dark matter particles could have been produced. Chung worked on a computer program to simulate the early universe and see if the rough calculations held up.

The results were surprising. Sure enough, you could make heavy dark matter particles this way. But the computer said you'd get the heavy particles even if the dark matter field didn't interact with the inflaton at all.

"We kept finding dark matter," Rocky recalled. "And we scratched our head for a day or a couple of days and said what the hell is this?" Maybe something was wrong with the computer program, they wondered. "Then we realized that it was in fact the gravitational production," Rocky said. The computer code contained the equations for the changing gravitational field. The math knew about the particle-production possibility that Schrödinger had identified in 1939.

About the same time, other researchers proposed similar ideas. So Rocky pursued the WIMPZILLA possibility more seriously. It turned out that if the young universe gave birth to WIMPZILLAS, they might have disintegrated into other particles by now. But it's possible that they disintegrate slowly—on a time scale comparable to the age of the universe. If so, enough of them may still be around to account for the dark matter.

He didn't stop there. In his relentless devotion to astronomical and lexicographical exploration, he also considered the possibility that WIMPZILLAS would actually interact strongly with each other. And if so they might better be known as SIMPZILLAS, short for strongly interacting WIMPZILLAS.

Whether WIMPZILLAS or SIMPZILLAS will solve the dark-matter mystery remains to be seen, of course. Plain old WIMPs may turn out to be all that astronomy needs. But that's not so obvious. WIMPs would be plain old cold dark matter, and for most of the past 20 years cold dark matter has been the favorite way to explain how galaxies formed and coagulated. But by the end of the century, a number of computer simulations began to cast doubt on whether the missing matter really is all cold and dark. Cold dark matter's problems were a hot topic at the Texas symposium in December 2000.

COLD AND FUZZY

Some physicists there suggested that the mystery mass is colder than cold, so cold that it becomes "fuzzy dark matter." Some argued that the dark matter is really lukewarm. Others suspected the need for yet another entirely new idea, generally described by the ugly acronym SIDM, for self-interacting dark matter.

"There's lots of crazy solutions out there," Ben Moore of Durham University in England said at the Texas symposium.

Moore reported there on what he called "the dark matter crisis," instigated by computer simulations of cold dark matter theory that did not produce pictures of the universe that look like the one we live in. Of course, the accuracy of the pictures depended on how closely you looked at them. On the grand scale of galactic clusters, the cold dark matter scenario looked pretty good. The simulations show that a universe full of cold dark matter should in fact produce huge clusters of galaxies, as astronomers observe. But the simulations don't show the right picture on smaller scales. Full-sized galaxies, like the Milky Way, should be surrounded by perhaps 1,000 smaller, dwarf "satellite" galaxies, the simulations show. But real-life surveys of the space around the Milky Way reveal only about 10 or so such satellites.

Other astrophysicists at the meeting pointed out further discrepancies. The density of matter observed in the cores of the dwarfs is less than the theory predicts, for example. And while the theory says the cores of the dwarfs should be lopsided, they actually appear to be pretty round.

All these observations seemed to say that WIMPs—generally regarded as the most likely component of cold dark matter—did not possess the proper properties to build the universe. And the culprit property seemed to be that WIMP particles did not interact very much with other particles or with themselves. (Remember, the *WI* part of WIMP stands for "weakly interacting.") Another way of saying this is that these particles don't bump into each other very much.

Princeton University physicist Paul Steinhardt says to think in terms of a pool table. If you spread a few very small spheres over the table—marbles, say—they can move around without much chance of colliding. But replace the marbles with billiard balls, and collisions become much more common. Assuming the problems with standard cold dark matter remain unsolved otherwise, Steinhardt said, it may be that the solution would require cold dark matter particles that collide with each other a lot.[14]

"The simplest explanation is that the dark matter is interacting with itself," he said at the 2000 Texas symposium.

Such "collisional" dark matter particles would interact often enough to deter the formation of dwarf galaxies around big ones, explaining why the Milky Way has only a few satellites. But the problem with collisional dark matter particles is that nobody knows what such a particle would be. Astronomers may need to be searching for something that hasn't even shown up yet in anybody's theory. "Perhaps . . . we should be looking for a different kind of particle altogether," Steinhardt said.

Rocky Kolb, of course, recommends WIMPZILLAS. "If you want strong interactions, then the WIMPZILLA scenario is very promising," he told me. (In that case the particles would be called SIMPZILLAS, as they would be strongly interacting.) On the other hand, Rocky and others have investigated the prospect that the simulation discrepancies might be solved with the one dark matter particle already known to exist, our friend the neutrino. Except for this purpose it would have to be a species of neutrino with more mass than scientists usually expect a neutrino to have. Extra mass would slow the neutrinos down, making them "warm" rather than very fast and "hot." Some "warm" dark matter might be the right ingredient needed to make the mix of galaxies, clusters, and satellites come out just right. (But it's not so clear that warm dark matter exists, and there remains some question whether it would solve all the universe's problems even if it did exist.)

Some of our other old friends might get involved in solving the dark matter puzzle as well. Vic Teplitz and Rabindra Mohapatra say that mirror-matter particles playing the role of cold dark matter would produce the observed density in the cores of dwarf galaxies. You never know.

I could go on and on describing possible strange candidate dark-matter particles. For example, some physicists suggest that the cosmic dark matter is actually a solid. This is very hard to picture, but it could be that the dark matter is distributed and evolves in such a way that the ordinary rules for describing solids would apply. Solid dark matter would have some pretty strange properties, of course, such as allowing people and planets and stars to move around through it.

An even more intriguing idea to me, perhaps because of its soft-and-cuddly sounding name, is known as "fuzzy dark matter." I believe the first use of this label came in a paper published in 2000 by three physicists then at the Institute for Advanced Study: Wayne Hu, Rennan Barkana, and Andrei Gruzinov. They calculated that extremely lightweight particles—it would take 10 million trillion trillion of them to weigh as much as an atom—might explain the structure of the dwarf galaxies and their scarcity. In the coldness of space, such minuscule particles would spread out in the form of waves, as dictated by the requirements of quantum theory. The waves of individual particles would overlap and merge, generating a fuzzy substance known as a Bose-Einstein condensate. Calculations show that fewer dwarfs might form from such a condensate, and the ones that did would be less dense at the core, as observations indicate.

Fuzzy dark matter particles sound similar to the axion, a particle proposed years ago by physicists Roberto Peccei, Helen Quinn, Frank Wilczek, and Steven Weinberg. (Wilczek named the particle after a brand of laundry detergent.) Axions, which solve a problem in the theory describing how the atomic nucleus holds itself together, would be very lightweight (though not as light as the fuzzy particles) but otherwise would behave much like WIMPs. And of course if axions exist, space might also be populated by axinos—their SUSY partners—another ingredient to consider in the strange mix of matters that might make up the universe.

On the other hand, the problems with cold dark matter may be resolved as observations (and theories) improve. The dark matter picture is always changing in some way or another. One year, everything seems fine with cold dark matter; the next, somebody has identified some observation that doesn't seem to fit the cold dark matter scenario. I can remember a talk from the early 1990s when one speaker displayed a cartoon with a tombstone proclaiming CDM-RIP. It came back to life.

In any event, the dark matter problem has proved to be both profound and difficult. The search for its identity goes beyond mere idle curiosity. Not only does it make up the bulk of the matter in the universe, but its properties determine why the universe is such an architectural masterpiece. Yet despite dark matter's importance, and all the attention paid to it for decades, its identity remains a mystery.

I suspect that the answer to the dark matter question remains elusive because theorists don't yet have as good a grasp on cosmology as they sometimes seem to believe. Clues to the answer, I think, may come with a better understanding of how the universe itself came to be—an understanding that began in the twentieth century with one of the most fantastic of prediscoveries: the expansion of the universe.

PART TWO

STRANGE FRONTIERS

THE BEST OF ALL POSSIBLE BUBBLES

5

From Friedmann and Cosmic Expansion to the Multiverse

Freshmen in college, exposed for the first time to academic rigor, often complain about the demands of their classes. I can still remember one of my classmates describing the difficulty of one professor's tests.

"His questions are all like 'define the universe, and give three examples,'" my classmate said. And in those days, it was an aptly ironic comment. Back then it was considered correct to say that if you've seen one universe, you've seen them all. Since the universe is all there is, there can be only one. It's common sense. A test question asking for three examples was asking the impossible.

Similarly, scientists used to say you shouldn't ask what happened before the big bang, the cosmic explosion that gave birth to the universe. It was a meaningless question, like asking who won the 1994 World Series. There was no World Series in 1994, and there was no time or space "before" the big bang—the very idea of "before time" makes no more sense than the idea of more than one universe.

Traditionally, cosmologists have scoffed at amateurs who dared to ask such questions. Imagine my fascination, then, to discover that some of those very cosmologists themselves secretly harbored the same wonders about what happened "before the beginning."

I remember when this realization hit me, while listening to the cosmologist Andrei Linde deliver a talk at a workshop in 1991, in which his discussion led up to asking what happened before the big bang. "It is impossible to ask the question," he said. "But it is impossible not to be curious about this."

After his talk, I interviewed him and asked for details. "I would say that what we are seeing now perhaps was not the big bang but was one in a sequence of bangs," he said. "There are many small bangs. The universe not only produces galaxies, it reproduces itself many times."[1]

So if Linde is right, it *does* make sense to ask what happened before *our* big bang. And it does make sense to talk about more than one "universe." The universe we see may be just one member of an extended ultracosmic family. As Martin Rees, the Astronomer Royal of Great Britain, puts it, "Our entire universe may be just one element— one atom, as it were—in an infinite ensemble: a cosmic archipelago."[2] It's as dramatic a shift in human thinking, Rees writes, as the Copernican revolution and the subsequent realization "that the Earth is orbiting a typical star on the edge of . . . just one galaxy among countless others." In a similar way, the universe may be just one "bubble" of space in a megafroth of cosmic carbonation extending far beyond the view of any conceivable telescope.

In a way, I shouldn't have been surprised when I heard Linde express these ideas—there had been whispers, after all, from time to time about such forbidden questions. In the 1980s, Linde and others had published papers describing a multiplicity of universes. And it was a logical enough deduction—I could have thought of it myself. I knew about inflation, the popular idea that in a split second after the universe was born, it had been disrupted by an enormous burst of ultrarapid expansion (as you might recall from Chapter 4). During inflation, one tiny patch of space almost instantly enlarged itself to a vastly larger scale, like a little wrinkled balloon transforming into the *Hindenburg*. (Or to be a little more precise in size, a pinch of space much smaller than an atom growing into the size of a basketball.) If some small patch of space inflated to make our universe, what happened to the rest of the space that didn't inflate? Was it just "out there," out of view? Well, maybe. But if one part of it ballooned to make our universe, why couldn't other parts of it blow up, too? For that matter, why couldn't any old patch of space, say, somewhere in New Jersey, decide to puff up into a new universe just for the fun of it? It would be an interesting twist for a future *Sopranos* episode.

Imagining all this would not have been possible, however, without imagining that the universe is expanding to begin with. That possibility seemed somehow beyond the scope of human imagination until Einstein, in 1917, realized that such a thing was thinkable. But Einstein didn't like the idea. The equations for his brand-new theory of gravity, general relativity, told him that the universe should be either getting bigger or getting smaller, when everybody knew that it was just sitting still, as it always had been. So Einstein "fixed" his equations, adding a term that made sure the universe stayed the same size for all time.

It seems that the first scientist to really appreciate the possibility, and to show in a concrete way that the universe may in fact be expanding, was the Russian mathematician Alexander Friedmann. And I think I know why Friedmann was the one. It had something to do with his taste in literature.

FRIEDMANN THE WEATHERMAN

In brief accounts of the history of cosmology, you will sometimes see a reference to Friedmann as a Russian meteorologist. I think that's how I first heard him described, and found it odd that a weatherman would have been concerned with general relativity. Later I came across an allusion to Friedmann the mathematician, and wondered whether that earlier reference to meteorologist had been a sloppy error. It turns out that Friedmann was both—a mathematician at heart, with weather forecasting as a day job.

Friedmann's story is compelling and tragic; he's an intriguing character who somehow seemed destined not to succeed but did anyway, sort of. His early death deprived the world of future prediscoveries, I suspect, and also limited the recognition he received for the main insight into nature that he did succeed in offering, the idea that the universe can expand. "He introduced both motion and development firmly in to the science of the Universe . . .," wrote his biographers, "and overcame and destroyed the centuries-old paradigm of the static nature of the Universe."[3]

He was born in 1888, son of a ballet dancer and composer (his father) and a pianist and music teacher (his mother). As a student in St. Petersburg, in the days before Lenin ruled the land and St. Petersburg became Leningrad, Friedmann was something of a student activist, involved in staging student strikes to protest government policies. He became a first-rate mathematician anyway, started in graduate school in 1910, and passed his master's exam in 1914.

But by then, he'd become intrigued with the weather, especially with applying math to the dynamics of the atmosphere. When World War I interrupted, Friedmann volunteered to help Russian pilots drop bombs on targets at the Austrian front. Applying his knowledge of math and the atmosphere, he developed some equations for predicting the proper release points to achieve the desired trajectories. "I have recently had a chance to verify my ideas during a flight over

Przemysl," Friedmann wrote to his friend Steklov in 1915. "The bombs turned out to be falling almost the way the theory predicts." In fact, goes the legend, the Germans somehow found out about Friedmann. When the Russian bombs were hitting their targets, German soldiers would say "Friedmann is in the air today."[4]

Friedmann wrote of several harrowing flights and dangerous landings, but he survived the war and came away with a deepened knowledge of the way math described how the atmosphere worked. In this way, I suspect, he learned to appreciate the physical meaning of his squiggles on paper. By figuring out how to relate mathematics to the physics of the air, he prepared himself to understand the connection between Einstein's equations and the behavior of the whole universe, of space and time.

In fact, as I read about Friedmann, I had concluded that it was this physical intuition for the meaning of math that made him the right person to realize that the universe can expand. And probably that did have something to do with it. Much later, though, I discovered a clue that hinted at another reason why Friedmann had such insight.

He was the first scientist, perhaps, but not really the first person to suggest that the universe expands. That idea dates back at least to the middle of the nineteenth century, and it came from the troubled but creative mind of the American poet Edgar Allan Poe.

EUREKA!

Although familiar with "The Raven" and "Annabel Lee," I'd never known that Poe had dabbled in cosmology. My first inkling of that came at a party in Santa Monica, given for me by my friends K. C. and Rosie[5] to celebrate the publication of my first book, *The Bit and the Pendulum*. That title, of course, alluded to Poe's short story "The Pit and the Pendulum."

Musical entertainment at the party was provided by a singer named Lynda Williams, known in the world of science as the Physics Chanteuse. Her songs evoke themes of primordial nucleosynthesis and the laws of motion. In discussing my book's title, she commented on the allusion to Poe and asked if he had been the one to solve Olbers' paradox—the mystery of why the sky is dark at night. I had no idea. An astrophysicist at the party, Kenneth Brecher of Boston University, said he didn't know either, but if Poe had discussed anything like that, it would be in an essay he wrote on cosmological issues called *Eureka*.

It wasn't hard to find references to *Eureka*, and I came across the full text on the World Wide Web. Sure enough, Poe had offered an explanation for Olbers' paradox, and much more. He envisioned objects that sound suspiciously like black holes, and described the universe as exploding outward from a point, expanding in size, and then contracting again.

Poe spoke of "the primordial particle," the embodiment of unity, that unity conferring upon it "infinite divisibility." "From the one Particle, as a center, let us suppose to be irradiated spherically—in all directions—to immeasurable but still to definite distances in the previously vacant space—a certain inexpressibly great yet limited number of unimaginably yet not infinitely minute atoms," Poe wrote.

Here was the big bang theory of the universe, in 1848, more than seven decades before Friedmann worked out the math. Talk about prediscovery. I was amazed. It was a great example of an imagination capable of discerning deep truths about existence, far in advance of their real discovery.

But of course, Poe was not a scientist. In retrospect, much that's right about the big bang can be seen in his writings, but some of it doesn't really fit so well. (The matter from his big bang blasted out into "previously vacant" space; in the modern view the space did not exist in advance of the big-bang explosion.) And he did not provide the critical ingredient needed to get credit for his foresight, namely, the mathematics that made it all quantitative and precise. It was just literary imagery mixed with physical philosophy, interpretable from a modern perspective as anticipating the work of Friedmann, perhaps, but not really related to it.

Except for one thing. Poe, it turns out, was one of Friedmann's favorite authors.

Not many people have made this connection. Among cosmological commentaries I have found several mentions of Poe's cosmological speculations, but never any mention of a connection to Friedmann. (I did find one obscure paper by a Poe scholar noting Friedmann's fondness for Poe.) And in truth, I don't know for certain that Friedmann read *Eureka* specifically or that if he did, it influenced his cosmological research. But I do know that Friedmann read Poe, from a passing mention in a Friedmann biography published in English in 1993.

One of the authors of Friedmann's biography, Viktor Frenkel, related a discussion in the small Russian town of Tim with Valentina Doinikova, a woman who had known the physicist Paul Ehrenfest. Ehrenfest, born in Austria in 1880, had married a Russian and spent time in St. Petersburg, where he knew Friedmann. So when Frenkel was interviewing Valentina about Ehrenfest, Friedmann's name came up. And it turned out that Valentina had at one time been engaged to Friedmann. So Frenkel asked about Friedmann, eliciting such information as that he usually wore a bowler hat, always carried an umbrella, and oh yes, his favorite authors included Dostoyevsky and Poe.[6]

It seems to me quite possible, then, that Friedmann was conditioned by Poe's imagination to see the true meaning of Einstein's equations, whereas others, Einstein included, did not.[7]

So it came to be that in 1922, Friedmann published "On the Curvature of Space," his first paper on relativity. "The purpose of this note is . . . to demonstrate the possibility of a world in which the curvature of space is independent of the three spatial coordinates but does depend on time," Friedmann wrote.[8] That is, in modern language, the size of the universe can change as time goes by.

In fact, Friedmann found within Einstein's equations the possibility that the universe could grow and then shrink. Perhaps, Friedmann suggested, the universe would grow and shrink just once. Or perhaps it could then grow again—an image of a periodic or oscillating universe that would have a beginning, an end, and then a new beginning. As Friedmann pointed out, the equations described possibilities but did not determine which of them the real universe actually chose. "Our knowledge is insufficient for a numerical comparison to decide which world is ours," he said.[9] But he noted that for a reasonable guess of the mass of the universe, its lifetime would be on the order of 10 billion years, not much off from current estimates of how long the universe has been around. What's significant, of course, is not the precision of that estimate, but that Friedmann thought in terms of a universe with possibly finite lifetime, rather than the everlasting cosmos that Einstein and almost everybody else believed in.

Einstein read Friedmann's paper but was unimpressed; he believed Friedmann had committed a mathematical error. In fact, Einstein had committed the error, and after two years of effort on Friedmann's part (with the help of a friend's visit to Einstein), Einstein relented and published an apology. (He still wasn't ready to admit that the universe expanded, however.)

Friedmann returned to the issue in a second major paper, published in 1924. In his first paper, he had treated the possibility that the universe was finite, meaning that the curvature of spacetime would be positive (like the curvature of the surface of a sphere). In his 1924 paper he explored the possibility that space was negatively curved (like the surface of a saddle). In that case the universe would expand forever and could be infinite in extent. (It would not necessarily be infinite, though, as we'll see in Chapter 9.)

For most of the rest of the twentieth century, cosmologists wondered which of the universes that Friedmann identified was the type we live in. But it was Friedmann who first clearly realized that those multiple possibilities existed.

I should mention that the science historian Helge Kragh has argued that Friedmann didn't really care about the possible physical manifestations of his universes but was only interested in the math.[10] Maybe that's right. But it sure seems to me that Friedmann must have believed his squiggles on paper could have physical meaning. He may not have known exactly what that meaning was, since the equations did allow different possibilities. But he did warn that interpreting the physical meaning of those equations required assumptions about the way that points in space are connected (the mathematical discipline known as topology). That also suggests to me that he did give some thought to his math's physical consequences.

In any event, it's impossible to know what Friedmann would have thought about the subsequent development of the big bang theory, or whether he would have contributed more himself, for he died in 1925. He had embarked on a record-setting balloon flight to record conditions in the upper atmosphere and overshot the landing zone, requiring an arduous trip back home. Shortly after his return, Friedmann became ill; the doctors diagnosed typhoid fever, and he died on September 16. (In the Friedmann biography, a relative alludes to his drinking some unboiled water on the return from the balloon trip.)

Though his career was cut short, Friedmann did live long enough to see—and to reveal to others—that Einstein's squiggles described a universe that need not be the static, unchanging, dull foreverlasting space of traditional science. Friedmann did, in fact, prediscover the expansion of the universe.

Credit for the actual discovery, of course, goes to Edwin Hubble. In his famous paper of 1929, Hubble deduced that galaxies fly apart from each other, and the farther apart they are to begin with, the faster they fly apart. The data he used to make that deduction were rather sketchy, but his interpretation survived over the decades as more and more observations were made.

The apparent explanation for the galaxies' higher velocities at greater distances was that the space between the galaxies was getting bigger. Hubble was very cautious in his original paper, though, couching the idea of expansion in technical cosmobabble. "The velocity-distance relation may represent the de Sitter effect . . . [including] a general tendency of material particles to scatter," he wrote.[11]

Translation of Hubble's discovery into the big bang theory of the universe's birth came from Georges Lemaître, a Belgian astrophysicist and clergyman, with later embellishments from George Gamow. (Gamow, in fact, had been one of Freidmann's students. I don't know if Gamow ever read Edgar Allan Poe.)

BEYOND THE BIG BANG

The big bang theory got its name in 1950, when Fred Hoyle, a British astronomer who didn't like the idea at all, used that phrase as a slur on a radio show.[12] It turned out to be a sound bite that came back to bite him; it gave the theory a vivid imagination-snatching label that made it easier to become popular. It made it possible for anybody, scientist or not, to think they had some understanding of how the universe began, even if they really didn't.

Most people, no doubt, think that some big explosion is all there is to the idea. There is little appreciation that the big bang theory is a complex mathematical framework, rooted in Einstein's theory of relativity and the equations that Friedmann and others developed from it. Still less do most people understand what cosmologists really mean when they say the universe began with a bang.

As it is most commonly articulated by cosmologists today, the standard big bang theory merely contends that long ago, probably

13 billion to 14 billion years before the present, the visible universe was very small, very hot, and very dense. That is to say, all of space and all the matter and energy it contains were wrapped up so compactly that ordinary ideas for describing space and time were rather meaningless. It's not the case that a small, hot egg burst into the space surrounding it; instead, space itself was confined inside the "egg." And then for some reason the tiny, hot, and dense space rapidly began to get bigger—that is, it exploded—everywhere at once. As this space grew bigger it cooled, eventually allowing familiar objects to form—first atoms, and then stars and galaxies, and ultimately planets and people.

So the picture of a gigantic explosion is a little misleading. In an ordinary explosion, some matter "blows up" and scatters itself into space. In the big-bang explosion that launched the universe, the matter and space were all tangled up together as they exploded into existence. In the standard view, there was no space, no time, and no matter "before" the big bang. The very idea of "before" the bang has no meaning, since there was no time until the big bang occurred.

Anyway, that's the standard view. But keep in mind, the idea of the big bang as an explosion is just a metaphor. Still, the evidence all points to a very hot, dense phase in the young universe, very much like the fireball of a big explosion. As the universe aged, it got bigger, just as material spreads out from the center of an explosion. So "big bang" really is a pretty good shorthand label.

For many years a competing view, championed by Hoyle and friends, was given equal space in many science textbooks. It was known as the steady-state universe. Its advocates accepted the expansion of the universe but could not believe it had begun suddenly one day with a big explosion. Instead, they surmised, somewhere in space matter was being continually created, out of nothing, in order to keep the overall appearance of the universe the same even as it grew bigger. To some, that scenario seemed as plausible as the big bang. But the steady-state view was slain in 1964, when Bell Labs scientists Arno Penzias and Robert Wilson detected the cool after-glow of the big-bang explosion, the cosmic microwave background radiation introduced in Chapter 4.

It is by now a well-known story: Penzias and Wilson wanted to study radio signals from space, but they could not eliminate all the sources of static from the antenna dish they were using as a radio telescope. Eliminating all possible sources of interference, including pigeon droppings on the satellite dish, left them with a faint noise from microwaves at a temperature barely above absolute zero. The microwaves came from all directions—no signal here from aliens or an enemy spy satellite. Somehow the background of space itself had a slight temperature.

Soon Penzias and Wilson learned from other physicists at nearby Princeton that the microwave radiation might be the afterglow of the big bang. (The Princeton scientists were making plans to search for the radiation themselves, but Penzias and Wilson beat them to it.) This cosmic microwave background radiation turned out to be the smoke left over from the big-bang gun. Objects throughout the universe today sit in a bath of radiation left over from the cosmic explosion that got the universe going to begin with.

Years earlier, Gamow had foreseen that the big bang should have generated such a relic radiation. He referred in a 1948 paper to the "high intensity radiation which remained from the original stage of expanding universe" before stars formed. Calculations of the radiation's temperature today were made by Gamow's collaborators Ralph Alpher and Robert Herman, who found that this microwave background should measure about 5° Kelvin, that is, 5 Celsius-sized degrees above absolute zero. And that turned out to be pretty close to the temperature that Penzias and Wilson found, 3.5° Kelvin, give or take a degree. (Today's best measurements give a temperature of 2.7°.) Considering the degree of difficulty in making this prediction, Alpher and Herman's result was phenomenal. Astrophysicist J. Richard Gott has compared it to predicting that a 50-foot flying saucer would land "on the White House lawn" and then watching as a 27-foot one actually does.[13]

Penzias and Wilson were not aware of the predictions, though. They heard about the explanation from Robert Dicke and colleagues at Princeton, who published a companion paper in the *Astrophysical Journal* providing the cosmic microwave explanation for Penzias and Wilson's data.

Even after those publications appeared, in 1965, many steady-state theorists refused to surrender. Still today sometimes they try to revive the steady-state corpse. But a consensus rapidly grew among most cosmologists that the big bang was the best idea to pursue for understanding the universe's origins. And from the mid-1960s on, several new developments strengthened the big bang's case. It would not take a Perry Mason to win a courtroom verdict in the theory's favor. The evidence is compelling:

Point 1: As Hubble showed, and subsequent work confirms, the universe is expanding. This is known because galaxies fly away from each other at a rate that depends on how far apart they are. (The farther apart, the faster they get even farther apart.) These speeds can be measured by shifts in the color of light emitted from the galaxy's stars. It gets redder the farther away a galaxy is, the way a train whistle or ambulance siren gets lower in pitch as it moves away from the listener.

The natural explanation for these observations of galaxies receding from one another in this way is that space itself is expanding—the universe is getting bigger every day. If the universe will be bigger tomorrow than it is today, it was smaller yesterday, and smaller still the day before that. Run the film of universal history backwards, and it looks as though the universe must have been a mere speck of its present self at some point in the past, roughly 14 billion years ago.

Point 2: The age of the universe, based on the expansion rate, is about the same as the age of the oldest objects within the universe as measured by other methods. If the universe had been around forever, it should have objects within it that are much older than 14 billion years, but it doesn't. Therefore the universe must have sprung into existence at some point in the past, as the big bang theory suggests.

Incidentally, the fact that the universe began at a specific time in the past and is now expanding explains Olbers' paradox about why the nighttime sky is dark. If the universe extended out infinitely, there would be enough stars to light up every point on the sky. So anybody claiming the big bang theory is wrong ought to keep quiet unless they can explain why it's dark at night.

Point 3: If there really was an explosion engulfing all of space, the radiation from that explosion should still be around, kind of like the glowing embers of a dead campfire. And sure enough, the faint glow of microwave radiation discovered by Penzias and Wilson is just such radiation. A hot primordial explosion explains this glow, as the initial heat would have prevented atoms from forming until the universe was perhaps 400,000 years old or so. Before then, photons of light would have mingled with the electrically charged particles of matter in an endless game of bumper cars, going nowhere. But as space expanded and cooled, the temperature eventually dropped enough to allow electrons to join with atomic nuclei, and the photons that had been bouncing around in the fireball would be free to stream through space. In the time since then, that hot radiation should have cooled to about 3° above absolute zero, just as the latest measurements have found.

Furthermore, if this picture is right, the radiation should arrive from all directions, with almost exactly the same temperature in all directions, and it does. And it should have a range of intensities at different wavelengths corresponding to what the big bang theory predicts. On all counts, the cosmic background radiation comes in right on the money. (By the way, this radiation causes some of the static you'd see on a TV tuned to a channel with no signal. If you eliminated every other possible source of that static, some would remain—further evidence for the big bang.)

Point 4: If the universe was as hot and dense as the equations of the big bang theory indicate, then the original soup of matter should have been cooked up into a specific mix of different chemical elements. The big bang theory can be used to calculate how much hydrogen, helium, and other light chemical elements should have been cooked up in that soup shortly after the universe was born. Those calculations suggest that the universe should be made of about three-fourths hydrogen, a little less than one-fourth helium, and traces of other elements. And in fact, observations of gases in space and the most pristine stars (uncontaminated by elements cooked up inside stars that later exploded) indicate that the amounts of hydrogen, helium, and other light elements are just what the big bang theory says they should be.

Point 5: Related calculations suggest that the neutrino, the subatomic particle predicted by Wolfgang Pauli, should be found in three distinct types, or flavors, but no more than three. Experiments at atom smashers independently find just that same limit on the number of neutrino flavors.

Point 6: If the big bang theory is right, it should explain why the universe is full of big clusters of galaxies. For those galaxies to be there, some little lumps of matter must have been around shortly after the beginning so they could pull more matter toward them by the force of gravity and in so doing grow bigger and bigger. If those lumps existed, they would have distorted the radiation of the big-bang explosion, producing little blips in the temperature of the microwaves found throughout space today. These very tiny temperature differences were detected by the COBE satellite in 1992.

While a few heretic astronomers have tried, nobody has succeeded in constructing any other reasonable theory that explains all these things. Explaining any one of the above points in any other way is hard enough. Explaining the whole package is something that only the big bang theory can do. Nothing else remotely comes close.

In other words, the big bang theory is not simply a plausible idea. It is a precise theory that explains, in a quantitative way, many features of the universe we see. It is hard to imagine that the ultimate theory of the universe will not incorporate the basic notions of the big bang in some form or another. And yet, the basic big bang cannot be the whole story. It explains a lot, but it doesn't explain everything.

WHAT HAPPENED BEFORE?

Again, in the standard view, there is no "before" the big bang. One way of looking at it is to think of the universe as beginning as a point—no size at all—where the matter-energy content of everything that would become the universe was infinitely dense. The universe began—time began—when that point started to grow.

It's hard to visualize three dimensions of space growing out of a point, but it's a little easier to imagine two dimensions growing in this way, as on the surface of the Earth. So let's just say the Earth represents the universe, keeping in mind that there's really an extra dimension we're not visualizing.

Start with a point, with all the matter and energy of the universe compressed to an infinitely high density. Then let the Earth's surface grow out from the point, curving away just as Earth really does curve away from the North Pole. As time goes by, the Earth gets bigger, growing ever southward. A trip around the Earth at its widest point keeps getting longer as time goes by.

Of course, the real Earth gets to a point of maximum width—the equator—and then starts to shrink again. If the universe is really like the Earth, it will expand only so much, reach a maximum size, and then get to be smaller again, like the Earth south of the equator. That is one possibility.

On the other hand, the universe could keep on getting bigger.

Instead of ending up like a sphere, the ever-expanding universe would look like an endless badminton bird, growing ever wider the farther down you went from the top. Or, the universe could keep expanding but at a slower and slower rate, eventually expanding so slowly that it would appear for all practical purposes to have stopped expanding. In that case the badminton bird would taper down to a nearly constant width.

Remember, this growth represents the passage of time. So what happens if you get down by the equator and decide to turn around and head north—the equivalent of going backward in time. Eventually you reach the North Pole and can go no farther. There is nothing north of the North Pole. And there is no "before" preceding the big bang.

This way of thinking has made most astrophysicists happy enough. Stephen Hawking (and his collaborator Jim Hartle) developed this idea fully into what they call the "no boundary" proposal—just as the Earth is round and has no boundary, the universe just "is," in a four-dimensional sense, with three dimensions of space and one of time. Time is merely one of the dimensions that is zero at the big bang the way the North Pole is 90° latitude. (We could just as well have called the North Pole 0° latitude, with the equator at 90° and the South Pole 180°.)

One way of viewing all this is to say that time is just an illusion based on the way our senses and brains perceive things. Everything, all of space and all of time, exists all at once, everywhere. We just move through different points on the time dimension the way a ship moves through the latitudes and longitudes on the ocean. When a ship is near the North Pole, the circumference of the Earth is very small; for anybody observing the universe at a time coordinate shortly after the big bang, the universe would be very small.

Personally, I don't find this view very satisfactory, but it does seem to conform pretty well to the picture of the universe that Einstein's theory of relativity provides, and it's a pretty good theory.

However, it is probably not a perfect theory. The assumption we made way back, about starting the universe out as a sizeless point with matter-energy of infinite density, is bogus. Remember, this theory is based on Einstein's equations of general relativity, and those equations break down when you get to a sizeless point. So we need a new, or at least modified, theory.

Sadly, nobody really knows exactly what that new theory is. There are many, many ideas about it, most beyond the realm of anything that can now be tested. But there has been one significant modification of the big bang theory that has achieved great observational successes and has offered as a bonus an astounding potential prediscovery: universes other than ours. The universe as we know it may not be the only such region of space that exists.

INFLATIONARY COSMOLOGY

It may be true that studying cosmology is a lot more expensive than it used to be, but that's not the origin of the theory of inflation. Inflation was born from the need to solve a few problems with the standard big bang theory. For example, on the largest scales, the matter in the universe seems to be spread out in a perfectly smooth way. It doesn't look like that to us, because we are very small compared to the universe.

If we look at a bucket of sand, it seems pretty much as though the sand grains are all smoothly distributed. But if we were very tiny little bugs crawling around in there, so tiny that one sand grain seemed huge, we'd notice that some of the grains are packed a little more tightly or spaced out a little more than others—in much the way that the galaxies out in space seem clumped a little more some places than others, with different-sized gaps between the clusters. From the point of view of someone who could get the big picture, however, the galaxies are spread out pretty uniformly. Besides, the microwave background is almost exactly the same temperature in all directions, further indicating that the universe at the beginning must have been a smoothly mixed soup.

But the original big bang theory cannot explain how the universe could get that smooth. It's as if something had thoroughly stirred it up at the beginning. (You could assume that everything just began all stirred up, but an assumption is not an explanation.) Cosmologists called this the "horizon problem." Think of the ordinary idea of a horizon—the point that represents as far as you can see. That means the point from as far away as light can reach your eyes. Translated to cosmology, the horizon indicates the distance that light can reach in the time available, or in other words, the distance that any physical influence can reach, since no physical influence can travel faster than light.

Even when the visible cosmos was very small, it was too big for light to travel all the way across it in the fraction of a second available before everything got blasted far apart. Yet the universe looks smooth. So what did the mixing? No physical influence could have mixed up the matter that was beyond the horizon in the early universe. A stirring rod would have had to be moving faster than light. That's the horizon problem.

Another conundrum unresolved by the big bang theory was known as the flatness problem. It referred directly to issues that Friedmann would have understood perfectly—namely, whether the universe would expand forever or not. Friedmann had worked out the math for various scenarios. If the amount of matter in the universe was sufficiently small, there wouldn't be much gravity to pull things together and the universe would expand forever. Too much matter, and gravity would eventually win over the expansion initiated by the big-bang explosion. The universe would someday stop expanding and begin to contract, eventually crushing everything back into a point—a scenario that eventually earned the nickname of "big crunch." Maybe the universe would then "bounce," and start expanding again, but that would be of no consolation to everyone who had been crushed to death. (Although that's not as bad as it sounds. As the astronomer Virginia Trimble once pointed out to me, the growing temperature as the universe collapsed would wipe everybody out long before the collapse was over. "You'd be fried before you'd get crunched," she said.)

There was, of course, a third possibility: that the universe would continue expanding, just at an ever slower rate, so that eventually it would be growing larger so slowly that for all practical purposes it would not be growing at all. And that scenario corresponded to an amount of matter that would make the geometry of space, on average, just like Euclidean geometry—the geometry of flatness.

In 1980 it was not possible to say for sure which scenario described the universe. But it was clear that the universe was pretty close to flat. It could not have very much less or very much more than the amount of matter that would place it on the borderline between eternal expansion and eventual contraction. And that realization posed a problem. For any slight variation from this critical density of matter would have been magnified immensely during the 14 billion years that the universe had been expanding. So at the beginning, the amount of matter must have been precisely fine-tuned to just the right amount.

But what physical process could have been responsible for that fine-tuning? Nobody had a clue. Except for Alan Guth.

In 1979, Guth was a young physicist at Cornell University, trying desperately to solve an obscure problem about a hypothetical particle that behaved like half a magnet. Now every child knows (or should know—based on some informal surveys, this isn't as widely known as it used to be), you can't have half a magnet. If you take a bar magnet and break it in half, the result is two magnets. Any magnet has two poles, designated north and south, and if you break it the new ends become new poles.

In the early universe, however, strange particles corresponding to only one magnetic pole might have been created. In fact, Guth calculated, huge numbers of these "magnetic monopoles" should have been created. But nobody had ever seen the slightest evidence of even one. So Guth and fellow Cornell postdoc Henry Tye tried to figure out why.

Developing an idea of the physicist John Preskill, Guth and Tye realized that the problem would go away if the process producing monopoles could be delayed somehow. Let's skip the details, except for one—Tye pointed out to Guth that the solution assumed that the expansion rate of the universe was not affected by the delay.

In December 1979, Guth began to explore that issue and shocked himself with the realization that the expansion rate of the universe *would* change because of the delay—dramatically. In a hundred billionth of a trillionth of a trillionth of a second, the universe would double in size 100 times. And a hundred doublings meant it ended up a million trillion trillion times bigger than it had started. Guth began referring to this tremendous burst of expansion as inflation, and he saw that it would solve both the monopole problem and the fine-tuning or flatness problem. Soon he discovered that his inflation also solved the horizon problem (although he had been unaware of that problem when he started). Such a rapid blast of expansion would have smoothed out any lumps just the way blowing up a balloon eliminates any wrinkles it possessed while uninflated.

Guth's original version of inflation was flawed in one critical respect—there was no apparent way for inflation to end. But soon an improved version of the theory came from Andrei Linde in the Soviet Union and from two young physicists in Pennsylvania, Paul Steinhardt and Andreas Albrecht.[14] Inflation rapidly became the most popular game in cosmology, with new versions popping up in the literature like new universes in the void. Within a decade, you couldn't tell the inflation theories apart without a scorecard—there was old inflation, new inflation, chaotic inflation, double inflation, hybrid inflation, and slow-rolling inflation. The proliferation of inflation versions suggested to some that the idea was wacky. But others believed that the basic idea was simply very robust—it worked so well that in whatever precise version of the big bang you liked, inflation of some sort would have to play an essential role.

Anyway, nowadays the evidence for inflation is very strong. Many precise observations of the microwave background, from satellites, balloons, and ground-based telescopes, have confirmed inflation's predictions. The inflated lady may not yet have sung, but most cosmologists believe inflation will win the game of explaining the way the visible universe looks today.

Some cosmologists, though, are not content just to explain the universe we see. They'd like to explore some of inflation's other ramifications—notably the possibility of other bubble universes that might have inflated somewhere out of sight. If such additional universes do in fact exist, they would surely represent one of the most amazing of prediscoveries. You could ask for no clearer example of discovering something before there was any physical evidence of its existence, because one of the properties of other universes is that there is no way to physically detect them.

UNIVERSES WITHOUT END

In this picture, Earth's universe lives in utter isolation from the others, one of countless separate bubbles of space somehow embedded in a vastly larger "metauniverse" or "multiverse." These other universes would be out of reach of any possible communication—no airline service, no e-mail, no message in a bottle could ever go there. And so, it would seem, all those far-off cousin universes are good for nothing. So why bother? Well, some cosmologists assert, maybe these bubbles *are* good for something. They may also help explain why the good old standard universe is the way it is.

Even with inflation, there are questions about the universe that astronomers still have trouble answering. Such as, why is it so big? And why do certain numbers from physics take the precise, constant value that they do? Why, for example, is the speed of light 300 million meters per second instead of 55 miles per hour?

One answer, popular with a few scientists but generally held in contempt, is that only a universe with certain features could contain people to ask such questions. Interest in this notion, known as the anthropic principle, began to grow in the 1970s, when British physicist Brandon Carter pointed out that small changes in many of the basic numerical constants of physics would render life impossible. He proposed a "weak" anthropic principle holding that the universe could not be much different and still support life. Other scientists proposed a "strong" anthropic principle suggesting that the universe has properties that require life to exist.

Most scientists think the weak form is pretty self-obvious, and the strong form is pretty self-stupid. Nevertheless, "anthropic" reasoning is alive and well in discussion of the possibility of multiple universes. After all, our "bubble" does possess properties hospitable to life. The question is, why? Perhaps the answer has something to do with the fact that it might be only one of many bubbles, all with different properties. Ours would naturally be the best of all possible bubbles—the one bubble with just the right mix of features to contain anybody to worry about it.

In this picture, the multiverse offers a bunch of different physics laboratories, each with different laws of nature. In some of those labs, under some of those laws, it is possible to create life. Assuming that the raw materials for making life must be cooked up in stars and spit out by supernova explosions, a very narrow range of values are possible for things like the masses of elementary particles and the strengths of elementary forces and other physical constants. Our universe has the physical constants it has, then, because it is the lab where life like us is possible.

Of course, many physicists reject the validity or the significance of the anthropic argument. Nevertheless, they still see in the equations of cosmology the possibility of multiple bubbles. And as these ideas have developed over the years, they've begun to exert influence outside of cosmology. They seem especially intriguing to anyone interested in the interface between science and religion.

UNIVERSES AND GOD

Multiple universes were a prime topic of discussion at a science-religion conclave I attended in 1999 in Washington at the Smithsonian's National Museum of Natural History.

Prominent scientists, theologians, philosophers, and historians gathered there to discuss the grandest of questions: Did the universe have a beginning? Was the universe designed? Are we alone? Among the participants was Guth, the originator of the inflation idea, now working on the details of what he called "eternal" inflation—the idea of a never-ending series of bangs producing new bubble universes.

Guth outlined the basics of inflation, telling how it essentially blew up a small patch of the whole universe into a bubble that became a universe unto itself. He reminded everyone that inflation solves some serious problems for cosmology, such as where matter comes from (it is created from the energy of empty space that drove the inflation) and why the universe scientists see today looks roughly the same in all directions (because it grew from one small homogeneous patch of space).

And if inflation could create one entire universe from a tiny patch of space, it could do it again, Guth said. And again. In fact, he asserted, countless other universes may continue to burst into existence—eternal inflation. If so, the question becomes whether the universe, or metaverse, or this series of bubbles, ever had a beginning, pushing the issue of creation of the cosmos into a new context. If the visible universe is just a bubble in a preexisting space, did the preexisting space have a beginning? Guth seeks the answer in his equations.

"It looks to me that probably the universe had to have a beginning," he said, "but I wouldn't bet everything on it."[15]

In either case, if eternal inflation is correct, the universe of today's textbooks is just one of countless many. To some of the theologians at the Washington conference, that sounded like bad news. It invites the conclusion that life exists simply because out of so many bubbles, one must be the Goldilocks universe—just the right temperature (plus other properties) for life to thrive. In other words, no specially designed habitat for humanity, just a lucky cosmic break. Instead of arising by design, human existence would be a happy coincidence.

In fact, some theologians try to argue that it's not "scientific" to propose all these other unobservable universes, and therefore creation of merely one universe by a god is a simpler—and therefore a more scientific—explanation. I find this reasoning to be rather transparently faulty. Scientists generally prefer the simpler of two *scientific* explanations, but a simple nonscientific explanation does not become scientific just because it is simple.

A more scientific (at first glance) objection, offered by some scientists, is that other universes are inherently unobservable. And that runs counter to the common scientific concept that what can't be observed is scientifically meaningless. In other words, whether other universes exist is more a philosophical question than a scientific one.

But many scientists argue that such a vision of science is much too limited. Scientists try to explain what they see, and sometimes the only explanation that works requires the existence of things that can't be seen. In other words, these extra universes might be a *necessary* part of whatever theory ultimately explains why our universe is the way it is. If a theory that explains everything that can be observed requires these universes to exist, then there would be a sound scientific basis for accepting their existence.

Martin Rees has made this point on many occasions. "The question 'Do other "universes" exist?' is one for scientists—it isn't just meta-physics," he writes. If a theory including numerous universes explains many hard-to-explain observable things, then that theory should be taken seriously. "If it predicts multiple universes we should take them seriously too, just as we give credence to what our current theories predict about quarks inside atoms, or the regions shrouded inside black holes."[16]

Science takes plenty of unobservable things seriously, he argues. For example, some galaxies may lie beyond the range of current tele-scopes, but nobody would argue, therefore, that those galaxies did not "really" exist. A new telescope might bring some of them into view tomorrow.

On the other hand, what about galaxies that are not visible be-cause they are beyond the horizon of space and time? Some galaxies are so far away that light traveling from them hasn't yet had time to reach Earth. No matter how powerful your telescope, you can't see them. But you might be able to see them someday—if the universe's expansion is slowing down, the day would eventually come when their light reached Earth. It would seem silly to say that they aren't "real" now but only will become real when their light arrives.

On the other hand, maybe the universe's expansion rate is not slowing down, but accelerating (thereby giving away the big surprise of Chapter 6). In that case, light from those too-distant galaxies will never reach Earth. They will never be seen. But does that make them less real than they would have been if the expansion rate were slower?

In the same vein, Rees contends that galaxies in some other uni-verse deserve to be just as real as those in our own bubble that we cannot see. Sure, these extra universes complicate the cosmic pic-ture, but they may very well be the necessary accoutrements of a cosmos that produces one universe that looks like the one we live in—with properties that allow us to live in it. (And, perhaps, with properties that allow us to prediscover things like unseen additional universes.)

"The multiverse concept might seem arcane, even by cosmological standards, but it affects how we weigh the observational evidence in some current debates," Rees notes. "Our universe doesn't seem to be quite as simple as it might have been. . . . Some theorists have a strong prior preference for the simplest universe and are upset by these developments. It now looks as though a craving for such simplicity will be disappointed."[17]

THE ESSENCE OF QUINTESSENCE

6

From Einstein's Greatest Mistake to the
Universe's Accelerating Expansion

*Quintessence is no other than a quality of which we cannot by our
reason find out the cause.*

—Montaigne

In a way, cosmology is like child's play.

After all, for decades astronomers have described the universe as
an expanding balloon. They knew it was getting bigger. They just
weren't sure whether it would keep growing forever or someday start
to deflate.

Now they're afraid that the balloon might burst.

In a plot twist worthy of *Scream 3*, the universe shocked cosmologists in the late 1990s with behavior almost as surprising as the original discovery that space is expanding. The new surprise was hyped
by the journal *Science* as the breakthrough of the year in 1998, even
though a lot of experts still didn't believe it. But by the time the

twenty-first century rolled around, the skeptics found it harder and harder to deny the new dogma: the universe-balloon is not only expanding, but it is inflating at a faster and faster rate. The best explanation seems to be that the universe is full of some mysterious cosmic fluid, unlike any known substance or force, utterly invisible, and therefore called "dark energy."

Nobody really knows what dark energy is or why the universe should be full of it. But if it really is out there, its existence should not have been all that surprising. It had, after all, been prediscovered.

"Yes, we anticipated it," the cosmologist Lawrence Krauss told me in an interview, "but I never believed it."[1]

Nowadays many people do believe in dark energy, even if it's too soon to say exactly which prediction it fulfills. Dark energy may represent the realization of an old prediction of Einstein's—one that he disavowed. Or it may signal a more recent possible prediscovery—a strange, variable energy field that has come to be known as quintessence.

Quintessence is not the name of some cosmic perfume. It is rather the revival of an old idea of Aristotle's. In his day, mainstream physicists believed earthly substance to consist of four elements (earth, air, fire, and water). But for the heavenly sphere of stars and planets, Aristotle invoked a surreal substance generally known as the ether. Later on, in Latin translation, this "fifth essence" came to be called *quinta essentia*.

Twenty-three centuries after Aristotle, Einstein's theory of relativity did away with the ether. But Einstein later found that the universe his theory described wanted to grow or shrink. So he added a term to his cosmic equations, known as the cosmological constant, which now threatens to revive something similar in many ways to the ether that he had done away with. In a sense, what Einstein took away, Einstein gave back.

EINSTEIN'S OTHER EQUATION

The cosmological constant story begins with Einstein's 1917 paper on cosmology, the first to apply the lessons of general relativity to the universe. It's in this paper that he explored the cosmological ramifications of what I call "Einstein's other equation." Almost everybody has heard of Einstein's first famous equation, $E = mc^2$; but very few outside the physics world could quote his other equation, $G_{\mu\nu} = -\kappa T_{\mu\nu}$. You can see why.[2]

So let's put it in words. In a nutshell the Einstein equation says that geometry of space is determined by all the sources of gravity in that space. (Technically, of course, we should be talking about the geometry of spacetime—space and time combined, as relativity requires. We'll deal with that distinction later.) The left side of the equation ($G_{\mu\nu}$) contains the symbol for representing how space is curved. The right side ($-\kappa T_{\mu\nu}$) contains the total of all the matter and energy (such as radiation) and pressure.

If the notion of curvature of space bothers you, don't worry, you're not alone. It is not the sort of thing that is easy to visualize. But you can see the connection between gravity and curvature if you think about something you can visualize, the surface of the Earth.

Earth is, more or less, a big ball, with a curved surface, and that curvature influences the path of objects trying to move in a "straight" line. Picture two ships at sea, steaming along on parallel paths, both heading straight north. As they near the North Pole, the ships would get closer and closer together. A bird flying overhead (let's call it Isaac) might remark that the ships appeared to be attracting themselves to each other. But then another bird (Albert) comes along and disagrees. No, says Albert, the ships get closer because the water they're moving on is curved. The curvature of the surface is moving them closer. In fact, Albert and Isaac might get into quite an argument over this, but they might conclude that the curvature of a surface and the attraction of two bodies seem to be just two ways of describing the same thing. In the end, though, Albert would be able to point out tiny differences in the ships' motion that differed from what Isaac would have expected.

In a similar way, Albert Einstein's theory of gravity makes subtle predictions that differ from Isaac Newton's. Objects move the way they do through space not because they are tugging on one another by way of an invisible gravitational force, but because space is curved—moving objects follow the curvature. Einstein's brilliant insight was that massive objects themselves cause the curvature of the space. A star or planet warps the space around it the way an obese gymnast distorts a trampoline. In the words of John Wheeler, "mass grips spacetime, telling it how to curve; spacetime grips mass, telling it how to move."[3] Einstein's other equation captures the mathematics of that insight.

When I first encountered that equation, I was baffled by the part about pressure. But I think it was because nobody ever bothered to tell me in school that pressure is part of the force of gravity. And no doubt nobody bothered because pressure rarely matters. But in principle, space's curvature at any location is affected not only by the density of matter sitting in it but also by the pressure exerted by any matter or radiation in the vicinity. Einstein's other equation takes all that into account.

Ordinarily, pressure is negligible. At the surface of the Earth, for example, the air pressure is trivial compared to the density of the mass in the Earth itself. And what really matters in curving space is the density of the energy, and to get the energy equivalent of a mass you multiply it by the speed of light squared. Consequently matter's mass-energy density contributes vastly more than its pressure to the curvature of space. Similarly, radiation pressure is usually pretty small—you can't lift an object against the force of gravity by shining a flashlight on it from below.

Under certain circumstances, though, pressure can be high

enough to be important—as in the center of a neutron star, where not only mass density but pressure as well is extremely high. Similarly, when the universe was very young and still a glowing fireball, the density of radiation was high and contributed a significant amount of pressure. In fact, in those days, the radiation-energy density exceeded the matter-energy density. Therefore cosmologists say that the universe was "radiation dominated" back then. But as the universe expanded, it cooled, and the radiation density diminished. Eventually the radiation's declining density dipped below the density of matter, and the universe became "matter dominated."[4]

One of Einstein's great realizations in the 1917 paper was that some third sort of component could compete for dominance with matter and radiation in affecting the curvature of space. In fact, Einstein believed he needed some such component to make sense of the universe. Without a new ingredient, the math told him, the universe would not stay still but would either expand or contract (depending on what initial conditions you applied the equations to).[5] Einstein, trapped in the paradigm of a static universe, and unaware of any evidence to the contrary, decided that the equations describing spacetime's geometry must be modified to prevent the universe from expanding or collapsing. He did not want the balloon to deflate or burst. So he added a fudge factor to his equation, thereby preserving a universe characterized by lack of character—an utterly changeless, finite universe in which the motions of stars were slow. This fudge factor, which came to be known as the cosmological constant, represented some sort of repulsion in space that prevented matter from collapsing because of gravitational attraction.

Einstein was vague about what this term actually represented physically. But later it was identified with some sort of energy residing in the vacuum of space. In other words, space without matter and radiation in it wasn't exactly empty. Some sort of energy inherent in the balloon's elastic kept it from changing in size.

A strange feature of this energy was the way it affected the pressure part of Einstein's other equation. The math said that this mystery component of the cosmos must exert negative pressure.

This idea can be very confusing, because it seemed like Einstein wanted a repulsive force to keep the universe from collapsing. You might think a positive pressure would exert a repulsive force. A negative pressure sounds more like a suction, pulling stuff in rather than pushing it out. But that's not the way Einstein's other equation works. Pressure (positive pressure) contributes to the attractive gravitational force. Negative pressure does the opposite, so it has the effect of repulsion. (Sometimes you see this effect referred to as "antigravity," but that doesn't seem quite proper, since pressure is a legitimate part of the gravitational equation. Negative pressure is just a part of gravity that affects the overall result in an unusual way.)

In any event, because positive pressure enhances gravitational attraction, negative pressure exerts a repulsive effect. So the natural tendency of attractive gravity to pull things together—and make the universe collapse—would be countered by the repulsive effects of a negative pressure, and the universe could maintain its static state.

Obviously there was no physical evidence whatsoever at the time for such an energy field, or substance, or fluid, or whatever it was. Not only that, it was not something that emerged naturally from Einstein's theory. He had to add it into his equation, simply because he thought the universe differed from what his original equation suggested. Einstein acknowledged as much when he noted that his added term was not required by anything known about gravity. "That term is necessary only for the purpose of making possible a quasi-static distribution of matter, as required by the fact of the small velocities of the stars," Einstein wrote.[6] In other words, since stars weren't flying toward or away from each other, spacetime must be static. (At the time, it wasn't yet clear that stars are clumped into galaxies.)

Einstein's use of a mistaken conception of the cosmos to deduce the existence of vacuum energy strikes me as a most unusual route to prediscovery. Dirac prediscovered antimatter because the equations told him to. In this case, Einstein seems to be telling the equations what to do. The theory itself did not require the addition of this fudge factor, but Einstein appended it to his equation anyway. Perhaps this is a good example to keep in mind in the effort to understand how prediscovery is possible.

In any event, Einstein's term was challenged a few years later by Friedmann, who (as perhaps you'll remember from Chapter 5) managed to escape the static universe dogma and establish the mathematical basis for contraction and expansion. Einstein was evidently not impressed, as he clearly retained belief in a static universe until the end of the 1920s. But then Hubble analyzed the motion of distant galaxies and found that they did move away from one another very rapidly—the farther away they got, the more rapidly they receded. The universe, it seemed, was not static after all.

Thereupon Einstein, according to George Gamow, called the cosmological constant the "biggest blunder he ever made in his life,"[7] a story retold perhaps more often than any other anecdote in the history of science. As far as I have been able to tell, there is no evidence for this assertion other than Gamow's remark in his autobiography, and Gamow's reputation for anecdotal accuracy is somewhere near the same level as Ronald Reagan's. But whether Einstein actually used the phrase "biggest blunder" or not, he did abandon the cosmological constant after Hubble's discovery.[8] In a later edition of his book on relativity, Einstein noted that he would never have invented the cosmological constant if Hubble's work had been done sooner. "If Hubble's expansion had been discovered at the time of the creation of the general theory of relativity, the cosmologic member would never have been introduced."[9]

DEAD, NOT BURIED

While Einstein attempted to bury his mistake, it was never quite forgotten. His original intent, to keep the universe safely static, no longer made any sense. But other cosmologists saw that the cosmological constant might still matter, even in an expanding universe. Students of quantum physics knew that even a vacuum could be full of energy, simply because quantum physics does not allow the energy of anything to be exactly zero. And while Einstein wanted vacuum energy to exert a repulsive force to keep the universe from contracting, such energy could just as easily alter the rate at which the universe is expanding. Vacuum energy might therefore influence measurements of the universe's age and size. In fact, an apparent mismatch between the universe's age and expansion rate was just the problem that many astronomers hoped that vacuum energy could solve. So the cosmological constant, the fudge factor in Einstein's equations representing vacuum energy, came back from the grave.

The age-expansion mismatch was a tough problem. For the universe, just as for a person, only certain combinations of size, age, weight, and growth rate make any sense. A 25-year-old man should weigh more than 40 pounds and not be rapidly growing, for example. But through much of the 1990s, the mass, expansion rate, and age of the universe seemed out of sync. Some measures of the expansion rate even indicated that the universe might be younger than the oldest stars, a rather blatant paradox. Adding Einstein's cosmological constant into the mix altered the calculations of the universe's age, resolving the discrepancy.

Vacuum energy could also solve the problem of the universe being "underweight." Most theorists believed that the geometry of space described by Einstein's other equation should conform to the standard Euclidean geometry taught in school. In other words, space would be flat, and two parallel light rays zipping through empty space would remain a constant distance apart. But Einstein's equation allows for space to be curved. Space might be curved like a ball (a "closed" universe with a lot of mass) or curved inward like a saddle (a lightweight, "open" universe). In a closed universe, parallel lines would converge, like longitude lines on the surface of the Earth. In an open universe, parallel lines would diverge.

In the simplest cases, the overall geometry of space would be determined by how much matter it contained. More than a certain "critical" amount, and the universe would be closed, eventually to collapse. Less than the critical amount, and the universe would be open and expand forever. A "borderline" universe, with precisely the right amount of mass, would expand forever, but at a slower and slower rate, and the geometry of space would be essentially flat. (Euclid's geometry would still be no good in the presence of a massive body, but overall, on average, space would seem to be Euclidean.)

If Alan Guth's inflation idea really did describe the birth of the universe correctly, we'd live in the borderline universe, and space's geometry should be very, very close to flat. If so, the universe must contain precisely the right amount of matter to bring the expansion almost to a halt. But various methods of measurement strongly suggested that the universe did not contain that much matter. Adding vacuum energy to the mix, though, could make up the difference. (Remember, all the energy adds up to affect space's curvature.) It seemed that if you wanted the universe to be flat, you needed to revive Einstein's cosmological constant, or something like it.

On the other hand, there was no real evidence that the universe was flat—that was more or less a theoretical prejudice. And the theoretical basis for vacuum energy didn't seem so sound, either. Calculations based on theory indicated that empty space should be teeming with an immense amount of vacuum energy—much more than observations allow. The equations predict the amount of vacuum energy to be higher than it actually is by a factor equal to 1 followed by 120 zeros. Nobody knows why the prediction is so far off.

"I think it's the biggest unsolved problem in cosmology today, and it's our biggest embarrassment," says Josh Frieman, a cosmologist at Fermilab.[10]

For a long time, most cosmologists favored the idea that there must be some reason why all the vacuum energy gets canceled out somehow, leaving zero. It was easy to imagine that some simple (but unknown) law of physics was at work, eliminating all the unwanted vacuum energy. It was not as easy to imagine that some mystery process would eliminate most of it, but not quite all. In short, the problem seemed too hard to solve. Therefore everybody went merrily on their way, assuming that the cosmological constant *must* be zero and therefore they could ignore it.

Along the way came some observations that provided comforting support for that point of view. I remember an astronomy meeting in 1992 where astronomers reported a study of the light from distant quasars using the Hubble Space Telescope. Quasars are the bright beacons on the universe's edges (from our point of view) shining beams of light across space like powerful cosmic flashlights. We can learn a lot about what's between the quasars and us by studying the light that they emit, because on its way to us the light is affected by what it passes through. Or around. Large masses, for example, exert gravitational force that bends the light as it goes by, the phenomenon known as gravitational lensing. A large mass could "lens" a distant quasar in such a way as to generate a separate image, so astronomers would see two images of the same quasar.

Now there's no point in going into the details, but vacuum energy would influence how many such double images you would see when surveying the sky for quasars. At the 1992 meeting, reports of such a study suggested that the amount of lensing was just what it should be, with no discernible effect from any vacuum energy. John

Bahcall, a prominent astrophysicist from the Institute for Advanced Study in Princeton, was emphatic. There might be a very small amount of vacuum energy, below the threshold of the telescope's capabilities to detect, but not enough to be cosmologically interesting. "It's too small to help you with any known astronomical puzzles," he said. "It's too small to be any good for anything."[11]

Other experts, however, said that this result was uncertain enough to allow different interpretations. So when I wrote about it, I suggested that it was unwise to jump to conclusions. "Don't be surprised," I wrote, "if the cosmological constant comes back from the dead again." And it did.

THE ACCELERATING UNIVERSE

Throughout the 1990s, cosmologists sought ways of solving their problems without adding dark energy into the cosmological mix. Some measurements of the expansion rate became more compatible with the ages of the oldest stars. And maybe, some experts suggested, the universe just didn't have enough matter to make space flat because space wasn't really flat. But as the twentieth century came to an end, new observations challenged all those attempts to avoid the need for dark energy.

Eventually, decisive evidence in favor of flatness came from the cold microwave background radiation, which contains an imprint of how the tiny seeds of matter were arranged in the very early universe. Taking the sky's temperature reveals the starting place for the evolution of the huge galaxies and galactic clusters that grew from those tiny seeds. Satellites, telescopes, and instruments on balloon missions can measure how much the radiation temperature differs between points separated by different angles on the sky. At some angles, the temperature difference is greater than at others.

The key issue is finding what angle has the greatest temperature difference. That peak difference depends on the influence of the cosmic equivalent of sound waves. In the early moments of the universe, gravity would compress radiation particles (or photons), but the pressure of the photons would eventually fight back and rebound. The compression-rebound cycle generates oscillations, or waves, that boost the temperature differences. The angle of maximum temperature difference can be calculated based on the assumption that space is flat. Any deviation from the expected angle would suggest that the universe is not flat after all. But by 2001, several experiments had measured the angle precisely enough to confirm that the universe was, in fact, very close to flat. So some sort of funny dark energy would seem necessary to explain everything else.

Even before the compelling evidence for flatness came in, another key advance had generated support for dark energy. That advance involved finding a way to measure the expansion rate of the universe in the distant past. If the universe in fact contained a lot of dark energy, the negative pressure would play havoc with the standard ideas of how the universe had expanded. For one thing, the universe would be expanding faster today than it was a billion years ago. But before the mid-1990s, astrophysicists had no good tool for measuring the universe's expansion history. Then they found a (standard) candle in the darkness.

At an astronomy meeting in San Antonio, in January of 1996, I first heard of a new effort to measure cosmic distances by exploiting the brightness of exploding stars, or supernovas. A particular version of supernova, known as Type 1a, seemed especially good for this purpose. In theory, they should all explode with the same brightness. Thus their distance could be inferred by how bright or dim they appeared.

Of course, it isn't that simple. In real life Type 1a supernovas do vary some in intrinsic brightness. And what's worse, distance alone doesn't make them dim. Dust gets in the way, too. A bright nearby supernova shrouded in dust might appear to be far, far away. And if you don't know how far away those objects are, using them to calculate the expansion rate is impossible.

But at the San Antonio meeting, scientists from the Harvard-Smithsonian Center for Astrophysics described a way to correct the calculations when dust interferes. The trick was watching how rapidly a supernova dims over time. An intrinsically bright supernova grows dim rather slowly; dimmer ones fade away more rapidly. And the dimmer ones tend to be redder than the bright ones, which are bluer. Dust's effect can be inferred by its effect on color. Dust blocks blue light but lets red light through, as in those reddish West Texas sunsets. By comparing the observed color with the expected color (based on how rapidly the star dims), the presence of dust can be detected and compensated for in the calculations.

At the same time, new computer-controlled telescope search strategies made it possible to track numerous supernovas in distant galaxies. So the race was on. Two separate teams embarked on such computer-aided supernova searches to find as many Type 1a's as possible. Two years later, the first results were in, and the shocking conclusion was revealed: it looked as though the universe really *was* expanding faster today than yesterday.

Many experts remained skeptical. But all of a sudden the idea of dark energy had to be taken seriously. In May 1998, dozens of leading figures in the field gathered at Fermilab for a workshop to discuss the implications of the accelerating universe, and dark energy became the topic of the day.

It is hard to relate the sense of excitement that flowed through that meeting, as though the cosmologists themselves were animated by a new form of energy. At a lunch with reporters, several of the scientists gushed over the new findings, like kids challenged to figure out how a wonderful new toy worked.

"We have no real clue for what this stuff is," said Josh Frieman.

"It's a monumental issue both for fundamental physics and for cosmology," said Paul Steinhardt.

"We were thrown a curve ball," said Michael Turner. "If it holds up, it's a surprise that the universe is accelerating rather than slowing down."[12]

In his talk at the workshop, Turner coined the name "funny energy" for whatever was in the vacuum and suggested that it was nothing other than the cosmological constant itself. "It was good enough for Einstein," Turner said. "It ought to be good enough for us."[13]

But if funny energy is the vacuum energy described by Einstein's cosmological constant, it suggests a curious coincidence. Vacuum energy arises from space in a roughly constant amount. As space stretches out, creating more space, more vacuum energy is created, too. So the density of vacuum energy in the universe remains constant at all times. For matter and other forms of energy, though, the situation is much different. There's only so much matter around, so as space expands, the density of matter diminishes.

Back in the universe's youth, the vacuum energy could not have been very significant. If the vacuum energy density exceeded the matter density from the outset, space would merely have blown apart and no chunks of matter would have been able to coalesce into stars or planets or anything else. Obviously, therefore, matter was denser than vacuum energy in the past. But now, the supernova observations suggest, the vacuum energy density is more important, driving an accelerating expansion. In cosmological terms, the switchover wasn't so long ago. Humans seem to have come upon the scene soon after the time that vacuum energy density and matter density were about the same. (In this case, "soon" means a few billion years, but for cosmologists that's like the day before yesterday.)

And so we have what Turner called the "Nancy Kerrigan problem." Why me? Why now?[14] In more technical language, it is called the cosmological coincidence problem.

At first glance, it does seem to be an amazing coincidence. Why should human astronomers be so lucky to be around at just the epoch when a changing number (the matter density) is just about equal to a constant number (the vacuum energy density)? At the Fermilab meeting, Steinhardt and colleagues suggested that the coincidence could be explained. Perhaps, they argued, the solution lies in returning not to Einstein, but to Aristotle. Maybe the dark energy is not Einstein's cosmological constant after all, but another mysterious energy form that they called quintessence.

QUINTESSENCE

Quintessence, supposedly, would be a "field," some sort of mysterious invisible fluid permeating all of space, like Aristotle's fifth essence. In fact, it made sense to call this new field a fifth essence as well. In Aristotle's day, remember, the standard model of ancient physics proposed that everything was made of four elements: earth, air, fire, and water. After two millennia or so of advances, physicists can still divide the matter and energy in the universe into four new categories: ordinary matter (made mostly of protons and neutrons); radiation; neutrinos (fast, light, ghostly particles); and cold dark matter, exact identity unknown. If there's something else—namely, dark energy—it seems logical to call it a fifth essence—or quintessence.

Steinhardt and colleagues coined the term quintessence in a 1998 paper, but the basic idea had been around for a while. Josh Frieman and Chris Hill, another Fermilab physicist, and collaborators had proposed something very much like it in 1995.[15] But nobody paid much attention until dark energy screamed out for an explanation.

Like the cosmological constant, quintessence would exert "negative pressure," kind of like the way a rubber sheet pulls in on itself. Or you could think of it as a stretched-out spring that wants to pull itself back in. Unlike the cosmological constant, though, it could interact with other stuff, and that property provides a possible explanation for the Nancy Kerrigan problem.

If quintessence exists, it fills all of space with some bizarre form of matter-energy that differs from Einstein's cosmological constant in an important respect—it isn't constant. Quintessence could solve the cosmological coincidence problem by being changeable; its strength could differ from place to place and time to time, being "thicker" in some places than others. "It's not like any other kind of matter that we're aware of," said Robert Caldwell, one of Steinhardt's collaborators.[16]

Of course, it's one thing to say that space is filled with energy that changes in strength over time in just the right way to solve all your problems. It would be much more satisfying scientifically to know *why* it changed in such a way as to solve your problems. The answer wasn't obvious at first, but by a year or so later Steinhardt had worked out an attempted explanation. The Nancy Kerrigan problem might be solved, he said, by something called "tracker fields."

WHAT IS A FIELD?

At the risk of getting a little technical, and with the cost of a brief digression, it helps to know that quintessence would be what physicists call a *scalar field*. And so it seems to me that the time has come to try to explain what a field is. Physicists throw the term *field* around as though they were talking about farms or baseball. But it isn't obvious to outsiders what the term *field* really means. Basically you can think of a field as something sitting in space—inseparable from space—that affects other things in space. In modern physics, almost everything comes from a field. Particles of matter are not tiny little balls, but rather are twists or knots in a matter field of some sort. But those fields are quantum fields, more complicated than the fields we need to talk about now. We can stick to classical fields.

Classical fields come in three flavors, scalar, vector, and tensor. Gravity (or, a gravitational field) is a tensor field. Multiple factors determine its magnitude, or strength, at any one point. A tensor field is actually just a more complicated version of a vector field, which has a strength and a direction at any point in space. Electromagnetic fields are vector fields—they have a strength and a direction at every point. If you put a little magnet at a point in a magnetic field it will orient itself in a specific direction. This is how a compass works, why the needle points north, because it is influenced to align itself in that direction by Earth's magnetic field.

Scalar fields, on the other hand, do not have a direction, just a magnitude. That is, a scalar field can merely be "thicker" at one place than another, or at one time or another. That's why the amount of energy residing in a quintessence scalar field can change over time.

Picturing the energy in a scalar field is not very easy, since scalar fields and energy are both invisible. But you can picture a graph of the amount of energy in a scalar field, and think about the field as representing an object moving along the graph. So let's picture the scalar field as a bowling ball rolling over an "energy terrain" and see if that helps.

Now all you need to know is the difference between potential energy and kinetic energy. The bowling ball's position on the terrain tells us its potential energy. On a high peak, potential energy is high, because as the ball rolls down the side of the hill it will pick up speed, and its potential energy will be transformed into kinetic energy, the energy of motion. If this bowling ball represents a scalar field, then the scalar field is thick on peaks and thin in valleys.

Now the natural tendency of a scalar field, just like a bowling ball, is to seek a state of zero energy—to give its energy up. Possessing a lot of energy tends to make you unstable, and nature abhors instability. Just as water runs downhill to the lowest point it can reach, energy fields seek their minimum possible value. The potential energy of the field (bowling ball) is just a measure of how far it is above the zero point it seeks.

The kinetic energy, on the other hand, is related to how fast the bowling ball moves toward the zero point. If it rolls downhill slowly—which means the strength of the field is changing slowly over time—its kinetic energy is low. If it drops rapidly, the kinetic energy is high.

The key point here is that kinetic energy makes positive pressure. A field that changes slowly has very little kinetic energy, too little to matter compared to its negative pressure. But a field that changes rapidly will have a lot of kinetic energy—more than its potential energy. If the kinetic energy exceeds the potential energy, the field will have positive pressure. That won't work with quintessence, which is supposed to have negative pressure. Therefore quintessence must be a scalar field that does not change very rapidly over time, so that it will have negative pressure. Because it must be changing over time very slowly to keep the pressure negative, quintessence is called a "slowly rolling" scalar field. It's like a bowling ball rolling down the energy slope very slowly to keep the kinetic energy small.

TRACKING QUINTESSENCE

If quintessence is a slowly rolling scalar field, it would mimic the effect of Einstein's cosmological constant. But that still doesn't explain why the magnitude of quintessence is just right to make the universe accelerate now, instead of a long time ago. At the Fermilab meeting in 1998, nobody had a good explanation.

But before long, Steinhardt and colleagues had cooked up a version of quintessence that had the potential to solve the problem. He described it at a meeting for science writers in Hershey, Pennsylvania, in 1999. Suppose, he said, that quintessence is a "tracker" field. In other words, quintessence interacts with other stuff in space in such a way that its strength can be suddenly changed when other things change.

For instance, it's plausible that when the universe was very young, quintessence interacted with the radiation in a way that made the quintessence strength decrease as the radiation density decreased. So as the universe expanded, and the radiation density dropped, so did the strength of the quintessence field. (In other words, quintessence "tracked" radiation.)

But then, at some point, the radiation density fell below the matter density. The universe became matter dominated. A tracker field might then respond in a different way—instead of continuing to fall in density, it might just stop wherever it was and then stay constant. But the matter density would continue to drop, eventually falling below the now-constant quintessence density. And the universe would begin to accelerate.

That could explain the Nancy Kerrigan coincidence. If matter domination triggers the quintessence field to stabilize, it makes sense that people would soon be around to talk about it, because matter domination also triggers the formation of galaxies, stars, and planets. It is no longer such a mysterious coincidence, but an obvious coincidence—the same event that triggers the processes needed to make people also makes dark energy take over the cosmos.

Of course, some serious details remain to be worked out, and quintessence has become a source of countless new research papers in the past couple of years. One especially interesting question is what quintessence implies for the future of the universe. If the dark energy is Einstein's cosmological constant, the universe's fate is sealed. For if it's constant, and will never diminish, once it dominates it will always dominate, and the universe will expand forever. But if the dark energy is quintessence, there is hope. If it changed in strength once, it could change again.

THE FATE OF THE UNIVERSE

Turner and Krauss explored this issue in a 1999 paper. The problem of the fate of the universe, they said, had been badly oversimplified in textbooks. For years astronomers had promised that the answer to the universe's fate was just around the corner. All they had to do was measure how much matter the universe contained. With a little help from some equations provided by Einstein, knowing the mass of the cosmos should have made it simple to say what its future held in store. Too much matter, and the universe would someday stop expanding and start shrinking, ultimately crushing itself—and everything in it—into nothingness. (The movie version of such a future would be called *The Big Crunch*.) Too little matter, on the other hand, and the universe would grow, cooling as it expanded, getting bigger and colder forever—more like *The Big Chill*.

More careful expositions pointed out that, technically, it's not the amount of matter that matters, but its density: A dense universe dies young, like James Dean; a not-so-dense universe just keeps getting older, like Strom Thurmond.

But dark energy changed the rules of the game. Density, Krauss and Turner proclaimed, does not determine destiny. Now it seems that the fate of the universe not only remains unknown, it may be forever unknowable. In other words, it's impossible to say how long forever will be.

If the dark energy is Einstein's cosmological constant, and is therefore always the same strength, it will inevitably exert more repulsive force than the attractive gravity of all the universe's matter. As the universe expands, the matter density naturally diminishes. Even if astronomers measured a matter density that seemed high enough to crunch the universe, they could never be sure that a tiny amount of cosmological constant might not be lurking in space below the threshold of detection. Vacuum energy paltry enough to

escape detection today would still dominate the universe in the distant future, ensuring endless expansion.

But if the dark energy is quintessence, it might not be constant at
all, but could change over time. In that case its future is unpredict-
able. "How do you know it's not going to disappear?" Turner said.[17]

Consequently today's measurements of matter density cannot
determine the universe's future. "Basically you may just as well throw
your hands up in the air," Turner said.

Even if it seemed the universe had far too little matter to collapse
(and that seems to be the case today), there are no guarantees. Just as
a tiny cosmological constant would someday take over and make the
universe expand forever, a tiny but negative vacuum energy would
someday cause expansion to stop and collapse to begin. The negative
energy would suck space in on itself.

Negative energy sounds bizarre, but it shows up frequently in
theory and sometimes in the lab. And a constant amount of negative
dark energy filling the universe—a "negative cosmological constant"—might very well be produced by various physical processes.
Physicists Je-An Gu and W-Y. P. Hwang of the National Taiwan University in Taipei made that point in a 2001 paper. If the universe
keeps expanding, any cosmological constant, positive or negative,
eventually becomes the most important factor in determining the
universe's fate, they pointed out.

"A negative cosmological constant, even if nearly zero and undetectable at present, can make the universe collapse eventually," the
physicists wrote.[18]

Krauss made the same point in his book *Atom*. "Once we acknowl-
edge the possibility that empty space can have energy, our ability to
unambiguously predict the future of the universe goes out the win-
dow," he wrote. "A negative energy in empty space could still result
in an extra-attractive force. . . . This would eventually stop the current expansion."[19]

"We really have no direct way of knowing the future," Krauss told me in an interview. No observation can be accurate enough to say for sure that a small amount of dark energy isn't hiding from view.

So scientists cannot say what the future of the universe will be. The universe may end in a hot crunch or a cold, endless expansion, and scientists will not be able to answer Robert Frost's question about fire or ice ahead of time.

Unless, of course, they discover the ultimate "theory of everything," which specifies precisely how much energy space must contain. "Having the ultimate theory is the only way we'll know," Krauss said. "I'm just not sure we'll have the ultimate theory."[20]

But many physicists believe that such an ultimate theory is almost at hand—if the universe turns out to be made of string.

SUPERSTRINGS

7

From Maxwell and Electromagnetic Waves to a World Made of Strings

It is a wonderful feeling to realize the unity of a complex of phenomena which, to immediate sensory perception, appear to be totally separate things.

—Albert Einstein

In the fictional world of Dr. Seuss, small is beautiful. "A person's a person, no matter how small," he wrote in *Horton Hears a Who*. But except for Olympic gymnasts, small gets no respect in real life. And sometimes not even in science. Which is why superstrings are often treated like the scientific equivalent of Rodney Dangerfield.

If they exist, superstrings are small in the extreme. A superstring would be smaller than a virus in the same proportion that a marble is smaller than the whole universe. To the critics, superstrings are superstitions. There's no evidence that they exist, and even if they did, they'd be so small that they could never be detected, and they are therefore meaningless figments of mathematical imagination.

Nevertheless, among many pretty smart scientists, superstrings are the most popular undiscovered objects in the history of theoretical physics. These supertiny loops might just be the ultimate stuff of all of nature's basic particles. Each species of particle might be a different mode of vibration of the fundamental superstring. That would explain the great diversity of particles in nature as merely separate manifestations of one primordial object. And so all the matter in the universe, everything from stars and atoms to green eggs and ham, might be made of string.

Best of all, superstrings offer a natural way to combine the math of gravity with quantum mechanics, paving the way to the unified theory of nature's forces and particles that Einstein dreamed of but failed to find.

THE SPIRIT OF UNIFICATION

Unification is a powerful motivation for physicists because they have seen such grand examples of its power. It's as though their motto is "one theory is better than two." By *unification*, physicists mean finding one explanation for different things, or finding one theory that incorporates others into a single unified mathematical package.

Newton, for example, established a whole new scientific worldview by unifying the physics of the heavens with the physics of Earth. Einstein superseded Newton by unifying space and time, matter and energy, and gravity and geometry with the theories of special and general relativity. And in between Newton and Einstein came another famous unification—the merger of electricity, magnetism, and light by the Scottish physicist James Clerk Maxwell.

Born in 1831, Maxwell was the son of a Scottish laird who enjoyed keeping up with the latest science and technology news, and had the resources to give his son a good education. Young James turned out to be fairly precocious. As a small child, on encountering a new mechanical device he would always ask "What's the go of that?" And when he received an oversimplified answer designed to satisfy a small child, he'd say. "But what's the *particular* go of that?"[1] Later he became a mathematical whiz kid, writing papers while still in his teens that were read before the Royal Society of Edinburgh. (He was too young to be allowed to read them himself.) He entered the University of Edinburgh in 1847, and after graduating went on to Cambridge, the alma mater of Newton, the greatest physicist of the seventeenth century. Maxwell became the greatest physicist of the nineteenth century.

Despite his unfortunately short life (he died at age 48), Maxwell was prolific. He mastered electricity and magnetism with a depth that enabled his equations to survive some of the assaults of twentieth-century physics that Newton's laws did not. Maxwell also developed the math for describing the molecular motion in gases, explained the rings of Saturn, and figured out the physics of color vision, inventing color photography on the side. It was his work on electricity and magnetism, though, which earned Maxwell his eternal reputation. He lived at just the right time to pull together the pieces of an electromagnetic picture that began to form in the early decades of the nineteenth century.

In 1820, Hans Christian Oersted in Denmark showed that an electric current could deflect a magnetic compass needle, an inescapable sign of a connection between electricity and magnetism. In 1831, the year of Maxwell's birth, Michael Faraday announced the discovery of electromagnetic induction—he could create an electric current by moving a magnet in the vicinity of a wire. Clearly electricity and magnetism shared not only some common features but also a deep and meaningful relationship.

From then on Faraday struggled to explain the nature of electricity and magnetism. Not much of a mathematician, he tried to understand electromagnetic relationships conceptually. He envisioned magnetic and electric lines of force that could extend through space in a way that would explain his experiments. Faraday created the first crude pictures of what nowadays is known as the electromagnetic field.

Along the way it became clear that light itself might get mingled into the new electromagnetic picture. In 1845 Faraday showed that a magnetic field can twist the orientation of a light wave. So the understanding of electromagnetism became entangled with old arguments about the nature of light extending back to Newton's time.

Actually, arguments about light had gone on far longer. But especially from Newton's time on, physicists had debated whether light was a wave or made of particles; nineteenth-century experiments weighed in heavily on the wave side. By mid-century no reasonable doubt remained; even Newton, champion of the particle picture, would have dropped his final appeal. He would have won a consolation prize, though, for the wave nature of light implied the existence of something else that Newton had suggested—the existence of an ether. If light was a wave, it needed a medium to wave in. You cannot, after all, have an ocean wave without water. And nobody could imagine a light wave without some "luminiferous" medium for light to vibrate in. It seemed natural enough to identify that medium with the old idea of an ether, some mysterious, invisible substance permeating all of space.

Of course, if there were an ether for light to wave through, there might be other ethers, too—perhaps one for electricity and magnetism. Faraday himself was skeptical of the ether idea, though. He thought his lines of force could exist on their own. Maxwell, however, sought to produce a thorough mathematical description of Faraday's ideas, and couldn't figure out how to do it without some space-filling medium.

So Maxwell tried to describe such a medium in a series of papers called "On Physical Lines of Force," published in 1861 and 1862. In the first paper he described the magnetic medium as a fluid of some sort consisting of magnetic whirlpools or "vortex tubes" arranged to correspond with Faraday's idea of magnetic lines of force. Maxwell developed the math to describe the stresses involved in the motion of the vortex tubes in a way that successfully delineated magnetic forces.

Working electricity into this picture was a little more difficult. In subsequent installments in his series of papers, Maxwell adopted a more elaborate model of his ether. He pictured the magnetic tubes as something like cylindrical cells separated by layers of particles, sort of like the steel balls used in bearings, to permit easier rotation. And a string of these cylindrical cells, surrounded by the bearing-particles, would then correspond to a magnetic field line.

More mathematical analysis showed Maxwell that if the cylinders spun at equal rates (corresponding to a constant magnetic field), the bearing-particles would spin but stay put. When nearby cylinders rotated differently—corresponding to a changing magnetic field— the bearing-particles would have to move through the medium. The math describing the particle motion in this model looked to Maxwell suspiciously similar to André Ampère's equations from the early nineteenth century relating electric currents to magnetic fields.

"It appears therefore, that according to our hypothesis, an electric current is represented by the transference of the movable particles interposed between the neighboring vortices," Maxwell wrote.[2]

There were further problems to resolve—such as how to represent static electrical charges—which Maxwell tackled later. Ultimately he succeeded, and "Maxwell's equations" describing electricity and magnetism remain among the most cherished creations in physical science.

Maxwell's success in turning Faraday's pictures into equations would surely have earned a Nobel Prize, if there had been Nobel

Prizes in those days. In any event, the mathematical unification of electricity and magnetism marked an enormous achievement in physics. But best of all, it came with a bonus—one of the greatest prediscoveries ever. For Maxwell had, in effect, anticipated radio waves. In fact, he essentially prediscovered the electromagnetic spectrum.

His original intent had been to show how the stresses in a medium—described mathematically—could account for electric and magnetic forces. In the process, he discovered that he could also account for the nature of light. For he realized that in his model of an electromagnetic medium, the charges and forces involved could produce disturbances, or waves, that propagated through the medium. It was simple enough to calculate how fast those waves would travel, based on the relative strengths of electric and magnetic forces. The answer came out to about 310 million meters per second.

Maxwell must have been pleased. For that number was almost exactly the speed of light. A famous 1849 experiment by Fizeau had measured light's velocity in air as 314,858,000 meters per second. In 1862, a new, more accurate experiment was conducted by Foucault, who reported 298 million meters per second. In either case, Maxwell's number was too close to be coincidence. (Nowadays, the speed of light is known to be just slightly less than 300 million meters per second.)

"We can scarcely avoid the inference that light consists in the transverse undulations of the same medium which is the cause of electric and magnetic phenomena," Maxwell declared.[3]

A couple of years later, Maxwell pursued the electromagnetic-light connection more deeply. He was not really happy with his mechanical model of the ether—the bearing-particles struck him as a little too artificial. In any case, he saw advantages in developing the math of his theory without depending on any detailed physical model. Rather than relying on the ether, Maxwell began to talk about the "electromagnetic field," and wrote in 1864 a paper called "A Dynamical Theory of the Electromagnetic Field," explaining that electrical bodies, "matter in motion," produced electromagnetic phenomena in the space around them.

This approach enabled Maxwell to describe light more specifically as a combination of electric and magnetic waves vibrating at right angles to each other. And it did not escape Maxwell's attention that such electromagnetic radiation might come in other forms. In one brief passage, he alluded to a source of many future discoveries: ". . . it seems we have strong reason to conclude that light itself (including radiant heat, *and other radiations if any*) is an electromagnetic disturbance in the form of waves propagated through the electromagnetic field according to electromagnetic laws."[4]

Nowadays we know that these "other radiations" do not require an ether, as Maxwell believed. But his belief in the ether wasn't critical to his conclusion. Although he seemed to believe in the ether, Maxwell didn't think you needed to know the details of the ether to describe electromagnetic phenomena successfully. Even so, I think it's significant that he arrived at his math by studying the relationships in a physical model to begin with. It reminds me of Alexander Friedmann, grasping the nature of the cosmos after learning to relate his squiggles on paper to the way the atmosphere behaved. Maxwell mastered the math of electromagnetism after working out the equations for describing rotating cylinders and balls. In both cases, finding math that represented physical relationships led to the pre-discovery of things not present in the physical models that produced the math.

Maxwell realized that nature did not have to use the same model as his to guide electromagnetic phenomena. But nature did appear to use a model that preserved the mathematical relationships that Maxwell had expressed. And those real relationships required the existence of previously unknown things: forms of radiation that nobody before Maxwell had imagined.

Once Maxwell clued others in to the possibility, though, the search was on. A few years after Maxwell's death in 1879, the German physicist Heinrich Hertz embarked on a series of experiments designed to test Maxwell's theory. By 1887 Hertz had detected hints of "electric waves" (nowadays we'd say radio waves) traveling through the air in his laboratory, and in 1888 he demonstrated their existence conclusively—confirming Maxwell's intuition and the power of his math to reveal previously hidden features of the physical world.

EINSTEIN, THE GREAT UNIFIER

Wrapping electricity, magnetism, and light into one neat mathematical package, Maxwell not only set the stage for radio and television, he inspired physicists who followed him with the notion that unifying nature's forces is a great way to make great discoveries. And nobody was better at playing the unification game than Albert Einstein.

Einstein was a simple man. He dressed simply, spoke simply, lived simply. There's a great story about Einstein using the same soap for shaving as for bathing. Discovering this, a visitor asked why he didn't use shaving soap. "Two soaps?" Einstein replied. "That is too complicated!"[5]

In physics, Einstein sought simplicity through the strategy of unification. "The real goal of my research has always been the simplification and unification of the system of theoretical physics," he wrote in his later years.[6]

His great early papers achieved unifications of enormous consequence for the future of physics. His special theory of relativity paved the way to merging space and time into a unified "spacetime." Another consequence of special relativity was that mass and energy are two sides of a unified coin.

Later, in his general theory of relativity, Einstein declared the equivalence of gravitation and accelerated motion and deduced that the geometry of space and gravitational force were also one and the same. His success with such unifications led him to seek the grandest unification imaginable at the time, the melding of gravity with the previous century's great unified theory, electromagnetism.

In pursuing that dream for the last three or four decades of his life, Einstein drew further and further away from the rest of physics. Yet even though he failed, Einstein's quest served an important purpose—as an inspiration for generations of physicists to follow. Einstein's followers soon realized that his failure reflected a deep schism between the twentieth century's greatest theoretical achievements—general relativity and quantum mechanics. Einstein's goal became reshaped. The grand challenge to grand unification, it seemed, would be reconciling general relativity with quantum theory. Only then, most experts reasoned, would it be possible to merge electromagnetism with gravity—and with nature's other forces.

For by mid-century, gravity and electromagnetism were not the only forces around to unify. By then the atomic nucleus had begun to liberate not only nuclear energy, but also some fundamental secrets of physics about where that energy came from. While Einstein's equation $E = mc^2$ could be used to calculate the release of energy from the nucleus, it didn't really explain it. The energy was available because particles within the nucleus are held so tightly together. Splitting a nucleus lessens some of the need for energy to do that binding, so energy can be released. But while they could figure out how to make a bomb by splitting a nucleus, physicists only dimly grasped the nature of the force that held the nucleus together.

Somewhat more progress had been made in understanding a second nuclear force—a force that causes some nuclei to decay radioactively and governs certain other features of nuclear behavior. This is the weak nuclear force—weak in comparison to the strong force that holds the particles together. (The weak force is responsible for the production of neutrinos, Pauli's prediscovery discussed in Chapter 4.) Part of Einstein's problem was his neglect of the nuclear forces. He believed they weren't so important and that if he succeeded in unifying gravity and electromagnetism, everything else would fall into place.

As it turns out, the first real progress came with the nuclear forces. In the 1960s, the weak force was shown to be very similar to electromagnetism. Proof came in 1983, when physicists at CERN found the W and Z bosons, the particles predicted by the math unifying electromagnetism and the weak force (remember Chapter 3). The strong force worked its way into the picture in the 1970s, as physicists developed the theory of quantum chromodynamics. By the mid-1970s physicists were fairly satisfied with their Standard Model incorporating the particles of matter and three of nature's forces.

But gravity resisted. Nobody could figure out a way to combine gravity with the other forces. Everything else fit together nicely, with quantum mechanics as the overarching framework. But gravity remained aloof, described exquisitely well by general relativity but sharing no common ground with quantum mechanics.

Meanwhile, a handful of theorists were in the process of pioneering an entirely new idea to explain the strong nuclear force. The new view envisioned the nuclear particles held together as though connected by a string.

STRENGTH IN STRINGS

At first, the physicists working on the problem of strongly interacting particles didn't realize they were dealing with strings. The math came before the idea. In particular Gabriele Veneziano, in 1968, found some interesting equations that described some strong-particle processes quite well. Other theorists developed further mathematical descriptions of the strong force at work.

According to John Schwarz of Caltech, one of the superstring pioneers, the early work had the look of "just a bunch of phenomenological formulas."[7] When it became clear that something substantial was emerging from the math, "it was natural to ask for a physical interpretation." As nearly as Schwarz can sort through the historical haze, three different physicists independently identified the physical idea at work: a one-dimensional object, or string.

One was Yoichiro Nambu, who mentioned the string idea at an obscure conference in 1969. Another was Leonard Susskind, who expressed the string concept in papers published in 1969 and 1970. In 1970, Holger Nielsen submitted a paper that also included the string idea to a conference in the Soviet Union. Then, in 1971, Pierre Ramond, André Neveu, and Schwarz developed a string theory variant that incorporated what turned out to be supersymmetry (which is why "string" theory eventually became known as "superstring" theory).

By 1974, though, string enthusiasm had diminished considerably. Quantum chromodynamics, string theory's rival for explaining the strong force, had made spectacular progress. Almost everybody came to believe then (and still does) that it is the "right" theory of the strong force. It fit in fine with the Standard Model. String theory, on the other hand, seemed, to be heading to a dead end.

Fortunately, John Schwarz was stubborn.

Schwarz had dabbled in string theory as an assistant professor at Princeton, and his work had impressed Murray Gell-Mann, the inventor of quarks. So Gell-Mann induced Schwarz to come to Caltech in 1972. "I didn't know what string theory was going to be good for." Gell-Mann told me. "I knew it was going to be good for something."[8]

So just as everyone else decided to cut all ties to string theory, Schwarz invited Joël Scherk to visit Caltech and work on strings. Scherk, a French physicist, had worked with Schwarz at Princeton. During the first half of 1974, they renewed their collaboration, exploring various aspects of string math, eventually focusing on a peculiar particle that kept popping up in the string equations. It was a particle without mass and with two units of spin, utterly unlike anything involved in the strong force. But Schwarz and Scherk realized that it did fit the description of one theoretical particle—the graviton, responsible for transmitting the force of gravity.

Soon Scherk and Schwarz were able to show that the stringy spin-2 particle would in fact behave exactly like the graviton. In other words, string theory contained gravity! Scherk and Schwarz had stumbled onto a quantum theory that was not only compatible with general relativity, it demanded general relativity. "Once we had digested the fact that string theory inevitably contains gravity we were very excited," Schwarz recalled. "Evidently, the way to make a consistent quantum theory of gravity is to posit that the fundamental entities are strings rather than point particles."[9]

And that wasn't all. Other features of the string equations looked just like the math used in the Standard Model for describing other forces. In other words, string theory turned out not to be a theory of the strong force but a theory of *all* the forces, and all the particles to boot. "This means that one is dealing with a unified quantum theory—an explicit realization of Einstein's dream," Schwarz explained.[10] So string theory might be the grandest of all unified theories—the "theory of everything" (a phrase, incidentally, that is detested by many physicists, including Schwarz and Gell-Mann).[11]

True, they had not actually devised a theory of everything, but Schwarz and Scherk had good reasons for supposing that string theory was the best foundation on which to build such a theory. And it seems to me like that should have been big news. But it wasn't, not even in the world of physics. Only a handful of physicists besides Scherk and Schwarz took strings seriously, and Scherk died in 1980. Most physicists still thought of strings as a theory of the strong force, and a better theory for that had come along. String theory also had another aesthetic problem, which we'll get to later. So for a decade, hardly anyone paid any attention to what Schwarz and Scherk had accomplished. But Schwarz stuck to his strings.

In 1979, he began collaborating with Michael Green of Queen Mary College in London. They explored the role of supersymmetry in the theory and made some progress, although slowly. Then in 1984, Schwarz and Green fired the shot heard 'round the superstring world, launching what later became known as the first superstring revolution. They produced a major breakthrough in the math, showing, in essence, how certain inconsistencies could be avoided. In particular, mixing quantum theory with gravity had always caused problems called "anomalies," in which symmetries related to conservation laws might be violated. And violating conservation laws is a bad sign. Schwarz and Green showed how in particular versions of superstring math, the anomalies would disappear.

For some reason, physicists took notice this time, perhaps because it seemed for the first time (to many of them) that superstrings might actually describe the real world.

All of a sudden string theory became the hottest candidate for theory-of-everything status. As news of Green and Schwarz's paper spread, other prominent physicists began to discuss the new results in seminars and colloquia. But word had not yet leaked out to the rest of the world. That was the situation until the spring of 1985. On May 7, a story appeared in the *New York Times* by the famous science writer Walter Sullivan. He began by writing:

> A number of leading physicists are beginning to suspect that everything in the universe is made of strings. Specifically, all of the basic particles of which the universe is made would be tiny strings—instead of points, as has long been assumed.[12]

This was big news to the general public, except for careful readers of the *Dallas Morning News*. For on April 22, two weeks earlier, I had written:

> The forces of nature seem to be neatly tied together by the theory of strings. . . . The idea behind string theory is that the subatomic particles that make up matter, long assumed to behave like points, really act more like strings.

In those days, I still had other duties at the *Morning News* besides science, and I couldn't keep up with every development as it happened. But I'd recently gone to a public lecture in Dallas by the famous physicist Steven Weinberg, of the University of Texas at Austin. Weinberg had won a Nobel Prize for his part in framing the unification of electromagnetism with the weak nuclear force, prediscovering the W and Z bosons found at CERN in 1983. He was widely regarded as one of the intellectual leaders in all of theoretical physics, the sort of guy that everybody took very seriously, as they do Edward Witten today. (Weinberg is still taken seriously too, by the way.)

As Weinberg talked, I realized he was describing a whole new twist on the frontiers of theoretical physics, a development that hadn't made it into the popular press yet. In fact, that is what caught my attention—Weinberg mentioned specifically that nothing had appeared yet in the news media about this development. Those are the words that every reporter wants to hear.

So I listened closely as Weinberg described the new idea: nature's particles aren't points, but strings. By *string*, he did not mean twine, but simply that particles should be construed as one-dimensional objects, like the line segments of Euclidean geometry.

Traditionally, I knew, physicists regarded particles as geometrical points—objects with no dimension at all. That was a necessary assumption, it seemed, because the math for describing particles did not work very well if you assumed them to be little marbles or something. In the quantum realm where subatomic particles roam, it is unwise to try to grasp things visually, anyway. It just turned out that for the purpose of calculating the properties of particles and describing their behavior, they acted like zero-dimensional objects, or points.

But the new developments, Weinberg said, showed that it was possible to describe particles as one-dimensional objects. And in fact, that approach offered some interesting advantages. He recounted the recent advance by Green and Schwarz and described how string math inherently included the graviton. That seemed to Weinberg to be a clear sign pointing the way toward the unification of gravity and quantum mechanics. String theory, Weinberg suggested, could be a "major breakthrough, possibly leading to another really great period in physics."[13]

It really did sound like an elegant idea. Weinberg discussed how the strings, small as they were, could be responsible for all the known particles of nature. By vibrating at different frequencies, or "notes," a string could mimic any elementary particle. The higher the frequency of vibration, the greater the mass of the particle, Weinberg explained.

And then he made an interesting point about the ultimate reality of the whole picture. "You can regard this as a mathematical artifice," he said, "or you can regard the string as having a real physical reality moving through spacetime. I'd like to downplay the reality of the strings."

So when I began working on a story on strings, I called John Schwarz at Caltech and raised just that point. "I know these strings can just be thought of as mathematical conveniences," I said, echoing Weinberg. "Oh no!" Schwarz exclaimed. "They're intended to be real, not just mathematical."

STRING FEVER

Over the next few years, more and more reporters wrote about strings, and more and more physicists jumped on the string bandwagon. And as more people worked on it, the theory got more complicated. Soon string theory split into five distinct versions, known as Type I, Type IIA, Type IIB, E8XE8 (also known as heterotic, or HE for short), and SO(32) (another heterotic version, known as HO for short). Each theory had its own mathematical peculiarities. Type I differed from the others in allowing two kinds of strings, "closed" and "open." Closed strings are like little loops, sort of like supertiny rubber bands. Open strings have unconnected ends, like a rubber band that has broken. While strings could be either open or closed in Type I theory, strings in the other four version are always closed loops.

String theory's five flavors cast suspicion over any claims that it was the obvious theory of everything. It seemed to many physicists that one theory of everything should be sufficient. The existence of five string theories provided further ammunition to the critics who didn't like the idea to begin with.

But the main reason superstrings got so little respect was their size, implying the impossibility of ever seeing them. Basic calculations showed that the natural size for the strings was so small as to be almost unimaginable. I've tried to express it different ways. It would take a trillion trillion of them to cross the smallest atom, for example. Or enlarging a superstring to the size of a real string is comparable to making a virus bigger than the entire Milky Way galaxy. If you want to know more specifically, the size of a superstring is something like 10^{-31} centimeters.[14]

On the one hand, the small size is good, because it explains why ordinary quantum theories, which treat particles as points, are so accurate. Experiments in the most powerful atom smashers might be able to probe sizes down to 10^{-16} centimeters. Particles (or strings) smaller than that would seem to behave like points, since the probe would not be able to reveal finer features. On the other hand, the small size is bad. An atom smasher with enough energy to probe down to the string size is beyond comprehension. With current technology, such a machine would have to be bigger than the solar system. So the tiny size of superstrings rendered it inconceivable that any microscope could ever "see" them. And thus critics broadcast the claim that it's meaningless to talk about superstrings to begin with.

Murray Gell-Mann does not agree. Sure, he says, the energy scale of superstring theories is too high to observe. "That doesn't mean that the theories have no consequences that are observable," he pointed out to me. The energy scale where phenomena related to the theory can be observed may be much lower than the energy scale where all the forces are unified.

"It's a really stupid mistake to mix those two things up," he said.[15]

Still, to some scientists, the impossibility of ever observing something makes it unscientific. But I think that point of view is overly simplistic, and reflects a fundamental misunderstanding of what science is and does. Sure, if you *define* what is real as what you can see, then you can reject a lot of things as meaningless. But if you regard science as a way of *finding out* what is real, and how the world works, then you may very well encounter aspects of reality that imply the existence of things you can't see. As Martin Rees argued about other universes, being unable to see them does not mean we cannot infer their existence from things we do see.

With superstrings, I think the case for their existence could be made even stronger than that for other universes. Because their small size implies only that strings would be very, very difficult to "see." That's a lot different from being impossible to detect in principle. When you argue that superstrings aren't real because no imaginable technology can detect them, you're basically saying that their unobservability is a practical problem, not that they're impossible to observe in principle.

I think this point is worth exploring a little because it gets to the heart of the relationship between math and reality. The math of Heisenberg's uncertainty principle, for instance, reveals something about nature. It shows that a particle can't have both a precise position and precise momentum at the same time. It is *not* merely that you can't measure them at the same time in practice—an electron does not possess these properties simultaneously, in principle. The theory does not allow it. With superstrings, the theory *requires* the existence of the strings. They are very hard to observe, but it's not impossible, in principle, to observe them, or their effects.

To me, the superstring critics sound suspiciously similar to the nineteenth century critics of atoms. The atomic theory became a central part of science in the nineteenth century, even though atoms could not be observed and some physicists contended that their "reality" was illusory. Even by the century's end, Ernst Mach, the physicist-philosopher who so greatly influenced the young Einstein, still denied the reality of atoms. Mach maintained to his death that atoms were mathematical fictions, unobservable and therefore not real. "Have you ever seen one?" he asked of anyone who disagreed.

To be fair, Mach was a truly deep thinker, and his analysis of the history of mechanics is one of the most insightful books about the nature of science ever written. Born in 1838, Mach began his scientific career only a few decades after the invention of modern atomic theory. While he was still a young professor, he developed a view of science as what he called "economy of thought." The world was a complicated place, with a lot going on. Nature presents all sorts of different phenomena. Science offers a way to categorize those phenomena and describe them simply and concisely. If you can boil down what goes on in nature to a few simple rules (call them "physical laws"), then you don't have to reevaluate every new situation you encounter in order to describe it or to predict what will happen next. That's economy of thought.

Atoms seemed at first glance to advance the economy-of-thought principle. It was possible to describe much of what happens in the world by assuming stuff is made up of tiny particles. But from his reading of scientific history, Mach concluded that all knowledge of nature was ultimately rooted in experience through the senses. So the proper province of science, he insisted, was understanding sensations. The primary features of the world to understand were colors, pressures, tones, and other sounds, the things that presented themselves to the senses directly. Atoms, of course, didn't do that. So for Mach, explaining phenomena in terms of atoms was just an analogy. To say the universe behaved in a mechanical way like a clock did not mean the universe was *really* a clock, Mach would have said. And to describe the world as made of atoms in motion did not mean that such invisibly small objects really existed.

"The mental artifice atom . . . is a product especially devised for the purpose in view," Mach wrote. "Atoms cannot be perceived by the senses . . . ; they are things of thought."[16] Someday, he believed, the "economy of thought" that science pursued would be possible without such fictional crutches; atoms were merely a temporary stopping place in the development of physical science. It would never be possible to detect them. "Atoms and molecules . . .," Mach wrote, "from their very nature can never be made the objects of sensuous contemplation."[17]

It took Einstein to find the flaw in Mach's approach. You did not need to make atoms accessible to "sensuous" contemplation to prove that they exist. In 1905, the same year that Einstein published his first paper on relativity theory, he found a way to show that atoms really do exist, beyond any reasonable doubt, even if it was still impossible to actually see one. The trick he used was to show how the visible motion of relatively large particles could prove the existence of invisibly small molecules.

The clue to Einstein's insight had been visible for decades, waiting around for someone to notice. When suspended in a liquid, particles large enough to be seen through an ordinary microscope bounce around like the Ping-Pong balls in a Lotto tank. That "Brownian motion" (first observed in 1828 by the biologist Robert Brown) could be explained only as the result of bombardment of the floating particles by molecules of the liquid, Einstein demonstrated mathematically. (In principle, there is no difference in this case between molecules and atoms.)

Although most scientists soon recognized that Einstein was right, Mach refused to yield. Until his death in 1916, he maintained his disbelief. Others might be willing to accept the indirect proof from Einstein's mathematics, but not Mach. He was sure that the existence of atoms could never be known, that atoms were forever out of the reach of the senses.

But he was wrong again, failing to foresee the possibilities of modern microscope technology. By the mid-1950s, a device called the field ion microscope succeeded in making pictures showing individual atoms (appearing as rather fuzzy dots of light, of course, but as it turns out, atoms are in fact rather fuzzy things). In 1990, IBM scientists, using an even fancier device called a scanning tunneling microscope, actually moved single atoms one by one to spell out *IBM* for a picture on the cover of the British scientific journal *Nature*. Mach would surely not have objected to using instruments to enhance the power of the senses. (He used special cameras, after all, to record the paths of bullets through fluids—research that led to the concept of Mach number for measuring the speed of sound.) So perhaps today even Mach would agree that the reality of atoms is well enough established to count as real science.

The story with superstrings may turn out to be similar. It might not be necessary to detect them directly—superstring theory could predict consequences accessible to smaller accelerators or other instruments. In fact, scientists today have already figured out ways that the existence of superstrings might be confirmed.

One possibility would exploit almost precisely the same trick that Einstein used to prove the existence of atoms. Physicists Ian Percival and Walter Strunz of the University of London have suggested that events on the atomic scale could be influenced by processes on the much smaller scale where superstrings operate, just as atoms and molecules bounce bigger particles around via Brownian motion. Since atoms behave like waves, devices could be built to measure interference patterns where individual atomic waves reinforce or cancel each other out. This process could detect properties of space and time on the scale of superstrings in just the way Brownian motion revealed to Einstein the properties of atoms and molecules.

In another approach, Massimo Galluccio, of the Astronomical Observatory in Rome, and his colleagues have studied the properties of gravity waves that may have been propagating through space since the earliest moments of the universe. Patterns in these waves could reveal signs of superstrings from the era when the universe itself was so small that the string size was significant. Gravity wave detectors now in place on the ground may not be able to detect such signals. But those effects might lie within the range accessible to a second generation of gravity wave detectors or perhaps orbiting gravity wave detectors (a project called LISA) that physicists have proposed.

So I think superstrings ought to be regarded as a prime candidate for prediscovery. Prospects for detecting them may very well be slim, but no slimmer than the prospects for detecting Pauli's neutrino.

In fact, the parallel between the neutrino and the superstring is intriguing in this regard. John Horgan's book *The End of Science* dismisses superstrings by noting it would take an accelerator 1,000 light-years around to provide evidence of their existence. That's ironically similar to the supposed need for a 1,000 light-year liquid hydrogen tank to detect neutrinos. The flaw in Horgan's reasoning, of course, is that there might be a way to detect superstrings other than by using an atom smasher. Just as the neutrino pessimists of the 1930s did not foresee nuclear reactors, today's superstring critics cannot know what invention of the future may offer a way, at least in principle, of testing the predictions of superstring theory.

On the other hand, there is one further objection to superstring theory that I haven't yet mentioned, another case where the mathematics implies an aspect of reality that many people find hard to swallow. For superstring math does not work unless space is a lot more complicated than almost anyone can imagine. Specifically, superstring math says space has more than 3 dimensions—maybe 9 or 10. The implication is obvious. Superstring theory implies that Rod Serling prediscovered The Twilight Zone.

PART THREE
STRANGE IDEAS

STRETCHING YOUR BRANE

8

From Schwarzschild and Black Holes to New Dimensions of Space

We are all liable to the same errors, all alike the Slaves of our respective Dimensional prejudices.

—Edwin Abbott
  Flatland

It now seems possible that we, the Earth and, indeed, the entire visible universe are stuck on a membrane in a higher-dimensional space, like dust particles that are trapped on a soap bubble.

—Steven Abel and John March-Russell,
  "The Search for Extra Dimensions"

Ordinarily it's unwise to admit religious evidence into scientific discussions. But let's live dangerously and consult the Bible.

In the King James Version, in John 14:2, we read:

In my Father's house are many mansions.

185

Biblical literalists might have a tough time with this passage. How could a "house" contain many *mansions*? Apparently, between King James's time and the twentieth century, the meaning of *mansion* changed. Originally it meant "dwelling place," so a house could contain more than one. So in the Revised Standard Version of the Bible, the verse reads "In my Father's house are many *rooms*."

That sounds better in light of the modern meaning of *mansion*. It makes no sense for something bigger than a house to be "in" a house. But in a way, the latest cosmological theorizing suggests that it is possible for something larger to be contained in something smaller. A whole universe, in fact, might fit into a space less than a millimeter wide. God's house may contain many universes.

It's a story like a mystery set in a mansion with a secret room, a room that contains many more mansions. As pop music's OMC would say, how bizarre, how bizarre. But it's very much like the story that many scientists have begun to tell about the universe. In what amounts to a real-life episode of *The Twilight Zone*, physicists have realized that nature may be concealing extra dimensions—not of sight or sound, but of space itself.

If so, the known universe may be just one of many "mansions" residing in the secret room of space's hidden dimensions. "It's just really frighteningly weird," says Rocky Kolb. "It strikingly flies in the face of everything we thought was true."[1]

On the other hand, it's a weirdness with a vast appeal to people who ordinarily find physics boring. Fermilab physicist Joe Lykken says, "This is the first thing I've worked on that my wife thinks is interesting."[2]

Normally, of course, dimensions themselves are boring. A dimension is basically just a number for describing position or motion. Any object's position in space can be identified with three numbers: latitude, longitude, and altitude. You can describe any movement using a mix of just three directions: forward-backward, left-right, up-down.

Space as we know it possesses three dimensions. (If you add time, the "fourth" dimension, you get spacetime.)

At first glance the world would make no sense with more (or fewer) dimensions of space. Yet contemplating extra dimensions, beyond the three of familiar experience, is not a new thing. Even some nineteenth-century scientists speculated on dimensions inaccessible to ordinary human senses. By the 1920s, some physicists realized that nature might actually have hidden away, at subatomic size, some extra dimensions too small to be noticed. By the 1980s, those tiny dimensions seemed necessary for the theory tying nature's particles and forces together with superstrings.

Physicists now wonder, though, whether the extra dimensions must really be so small, vastly smaller than an atom. Maybe those dimensions are as big as a millimeter across—10 million times the size of an atom. A dimension like that could be just like the mansion's secret room—in one sense smaller than the mansion, but in another sense big enough to contain many additional mansions.

Apart from evoking the science-fiction fantasy of parallel universes, this view of space offers possible solutions to several cosmic problems. The hidden dimensions may contain clues about the nature of gravity, the origin of the universe, and the identity of the dark matter. If these hidden dimensions exist, they would represent another astonishing prediscovery, validating the prescient imagination of *Flatland* author Edwin Abbott and the mathematical insight of such scientists as Savas Dimopoulos and Lisa Randall.

Hidden dimensions would show once again that space is stranger than fiction, much the same way as another famous prediscovery revealed a place to go beyond sight and sound—the black hole. And in fact, the story of black holes ends up to be tied directly to the extra dimensions required by superstrings.

VISIONS OF BLACK HOLES

Black holes have a long history. After they were first imagined, it was almost two centuries before they were named and more than two decades after that before they were found. In its original form, the black hole idea emerged from Newtonian physics, perhaps another example where a good prediction was based on errant premises. Nevertheless, I think it's fair to credit some insight to the man who first proposed the existence of stars so dense that light could not escape from them. His name was John Michell, born in England in 1724. He started his career as a geologist at Cambridge, where he was an expert on earthquakes. At age 40 he became rector at Thornhill in Yorkshire, but he kept his hand in science, turning his interest to the stars, and to gravity.

In 1784, Michell published a study on what you could find out about stars by examining the light they emitted—their distances, sizes and masses, for example.[3] In those days, light was not understood very well, so drawing grand conclusions from studying it was a risky proposition. For one thing, Michell believed, like all good Newtonians, that light consisted of particles. So it was natural for him to imagine that a star's gravity should have some effect on the particles of light that the star emitted.

Suppose a star, or some other body, was 500 times wider than the sun, yet just as dense. Its mass would be enormous, and its gravity, therefore, very strong. Michell calculated that light particles do not fly fast enough to escape the gravity of such massive objects. "Their light could not arrive at us . . . we could have no information from sight," he wrote.[4]

A few years later, the French mathematician Pierre Simon de Laplace performed a similar calculation. Other speculations about black-hole-like objects appeared in the nineteenth century—Edgar Allan Poe, for example, described something like one in his prose poem *Eureka*. But the first modern math hinting at a black hole's possible existence came from a German astronomer whose name in English means black sign—Karl Schwarzschild.

Born in 1873 at Frankfurt am Main, he showed exceptional mathematical ability as a teenager and went to school at Strasbourg and Munich, where he applied his math skills to astronomy. He became a professor at the University of Göttingen in 1901. By 1909 he had become director of the astrophysical observatory at Potsdam.

Like Michell, Schwarzschild was very interested in light from stars, and he developed new techniques for observing and analyzing it. He also developed an interest in grander themes, such as the geometry of space itself—an important point to be discussed in Chapter 9.

But his fame today derives from work he did shortly before he died, from illness he contracted as a soldier on the Russian front during World War I. Even during wartime Schwarzschild tried to keep up with science, and late in 1915 he was intrigued by a new report in the *Proceedings of the Prussian Academy of Sciences*. It was a paper by Einstein—the first publication presenting the brand-new general theory of relativity. Einstein's paper set out the basic equations of the theory. It was not, however, a trivial matter to solve the equations when they were applied to a specific real-world situation of any complexity. Schwarzschild, though, immediately saw a way to solve the equations in a relatively simple situation, the case of a perfectly round star.

Schwarzschild used Einstein's equations to calculate the gravitational field around a mass point, which would describe the gravitational field inside a star (assuming that the star was not spinning). In a second paper he calculated the geometry of the spacetime around such a star—in other words, how a star of a given mass would warp the spacetime around it.

Schwarzschild zipped off the two papers to Einstein, a few weeks apart—the first one on the mass point, the second describing the space outside the star (in Schwarzschild's terms, the gravity of a "sphere of incompressible fluid"). Einstein reported the findings to the Prussian Academy in early 1916. By May of that year Schwarzschild was gravely ill with pemphigus, an incurable skin disease. He returned home to Potsdam but died on May 16, and his exploration of Einstein's new universe was over.

Eventually, though, Schwarzschild's solutions would become immensely important for later explorers. "The Schwarzschild geometry," the physicist Kip Thorne has written, "was destined to have enormous impact on our understanding of gravity and the universe."[5]

In Schwarzschild's solutions, a curious quantity appeared: a distance from the center of a star at which the equations suddenly broke down. (As Jeremy Bernstein has put it, "the mathematics goes berserk . . . time vanishes, and space becomes infinite."[6]) Such a strange mathematical quirk would seem to demand some explanation of what its physical meaning was. For Schwarzschild, apparently, the answer was that a sphere could not get smaller than this radius, at least as measured by "an observer measuring from outside." This distance from the center (or "Schwarzschild radius") depends on the mass of the object.

For a sphere the mass of the sun, Schwarzschild noted, the distance would be 3 kilometers, about 2 miles. It seemed to Schwarzschild that for any known star, the critical radius would always be inside the star. It would be impossible, he calculated, to reduce a spherical incompressible fluid to smaller than that size; as you squeezed it down the pressure would become infinitely great a little before you reached the Schwarzschild radius.[7]

Once again, though, history teaches the foolishness of ignoring mathematics' power to reveal new phenomena in nature. Schwarzschild, without knowing it, had provided the fundamental insight leading to the discovery of black holes.

NEW BLACK HOLE HORIZONS

Still, black holes are so bizarre that it took decades before the physics community was ready to ponder their actual existence. Einstein himself considered the possibility of black holes and rejected them. In a 1939 paper, he calculated that any group of objects (say, stars in a cluster) packing close enough together to approach the Schwarzschild radius would become unstable. The objects would have to begin moving faster than the speed of light at such short distances—an impossibility, of course, according to Einstein's special theory of relativity. Therefore, no black hole could form. "The essential result of this investigation is a clear understanding as to why 'Schwarzschild singularities' [that is, black holes] do not exist in physical reality," Einstein wrote.[8]

But in the same year, another paper appeared, this one arguing in the opposite direction. J. Robert Oppenheimer, soon to lead the Manhattan Project, and his student Hartland Snyder, were exploring the physics of collapsed stars. Already, Subrahmanyan Chandrasekhar had calculated his famous limit on how massive a white dwarf could be—about 1.4 times as massive as the sun. A white dwarf is a burned out sun, no longer producing enough energy to counter the inward pull of its gravity. With no pressure pushing outward, a white dwarf would shrink down to dwarfdom, until the pressure of compressing the subatomic particles themselves came to the rescue and prevented further shrinkage. But for a star with a mass of more than 1.4 suns, the gravity would overwhelm even the subatomic particle pressure.

A stellar remnant with too much mass might solve the problem by exploding, leaving behind a neutron star, in which the subatomic particles have, in essence, merged to create neutron matter. But even then, neutron stars have a mass limit that isn't all that much greater than for white dwarfs. So it seemed to Oppenheimer that a sufficiently massive star might very well shrink down to a size of less than the Schwarzschild radius. He wondered what would happen then.

Apparently unaware of Einstein's paper, Oppenheimer and Snyder produced a surprising conclusion—the answer depends on your point of view. To an observer far away from the collapsing star, its light gets redder and redder. The increasing strength of gravity— or warping of spacetime around the star—slows time down, stretching the star's light to longer and longer wavelengths (redder and redder colors). At the Schwarzschild radius—remember, time is frozen—it stops altogether, so the light doesn't even leave. It is frozen in time, and a distant observer sees no more action from the collapsing star.

Snyder and Oppenheimer reasoned that pressure would resist the continuing collapse of a star as it approached the Schwarzschild radius. But to simplify their calculations, they supposed there was no pressure. Then Einstein's equations provided no solutions that could describe a situation in which the star stopped shrinking. "When the pressure vanishes . . . we have the free gravitational collapse of the matter," Snyder and Oppenheimer wrote. "We believe that the general features of the solution obtained in this way give a valid indication even for the case that the pressure is not zero, provided that the mass is great enough to cause collapse."[9] In other words, with a star massive enough, the collapse caused by gravity would overwhelm whatever pressure there was.

And that gravitational collapse would create, in the eyes of someone watching from far away, just what we now call a black hole. As Snyder and Oppenheimer put it, "the star thus tends to close itself off from any communication with a distant observer; only its gravitational field persists."

For a nearby observer, however, nothing is frozen, and the situation would appear quite different. "Near the surface of the star . . . we should expect to have a local observer see matter falling inward with a velocity very close to that of light," the physicists wrote.[10] Any such observer might fall in as well, winning the chance of a lifetime to explore the arena within the Schwarzschild radius. But it would be a short lifetime. Once inside, any object, observer or otherwise, would be drawn to the intense gravitational attraction at the center, getting ripped to shreds on the way in—although Oppenheimer and Snyder didn't mention that problem. They did calculate that an observer riding along with the infalling mass of the star would soon be unable to send a light signal to the outside world, and that "this behavior will be realized by all collapsing stars which cannot end in a stable stationary state."[11]

Although this paper basically described the black hole picture pretty clearly, it seems that nobody paid much attention to it. Even when Oppenheimer died, in 1967, it wasn't discussed in the obituary written by Hans Bethe for the Royal Society. Of course, the obvious reason for this lack of attention was that the paper appeared in *The Physical Review* on September 1, 1939, the day World War II began.[12]

No doubt the intervention of the war diverted everybody's attention elsewhere—especially Oppenheimer's. And maybe in those days science-fiction films hadn't been imaginative enough to prepare science for such strange possibilities. In any event, it was almost two decades before anyone seriously took the issue up again.

In the late 1950s, John Archibald Wheeler at Princeton had been reviving interest in general relativity, beginning to produce a stream of outstanding students who would dominate the field for the rest of the century. Soon issues involving the Schwarzschild radius arose, but at first Wheeler was skeptical about the possibility of gravitational collapse below that limit. At a 1958 meeting in Brussels, Wheeler challenged the Oppenheimer-Snyder result. The weirdness of gravitational collapse into nothingness must somehow be eluded in real life, Wheeler believed. But he had no solution to the mystery of what happens to very large masses undergoing collapse. Oppenheimer was there, and he defended his original conclusion. "Would not the simplest assumption be that such masses undergo continued gravitational contraction and ultimately cut themselves off more and more from the rest of the universe?" Oppenheimer asked.[13] Wheeler didn't think so then, but he would later.

In December 1963, the situation began to change. At a famous meeting in Dallas—the first Texas Symposium on Relativistic Astrophysics—the Oppenheimer-Snyder paper was a hot topic of discussion. In the next few years, Wheeler's view changed, and the discovery of pulsars in 1967 dramatically demanded a better understanding of gravitational collapse. Wheeler came not only to believe in black holes, he even christened them. I told the story of how they got their name in a brief account in the *Dallas Morning News* in 1998:

> In 1967, the discovery of dense, pulsating stars known as pulsars sparked further interest in the fate of heavy stars. Wheeler discussed the issues at a conference held that fall in the wake of the pulsar discovery.
>
> After he'd used the phrase "gravitationally completely collapsed object" several times, someone in the crowd—Wheeler still doesn't know who—offered a suggestion.
>
> "Somebody in the audience piped up, 'Why not call it a black hole?'" Wheeler said in an interview. . . .
>
> Wheeler liked the suggestion and decided to slip the term into a talk at the end of 1967 that was published the following year, the first official use in print of the astrophysical "black hole. . . ."[14]

Still, naming them wasn't the same thing as finding them. Though their theoretical possibility had been established, scientists debated for another quarter century whether black holes really existed. Several strong candidates had been discovered in the 1970s and 1980s, but loopholes always existed, in the form of possible alternate explanations or uncertainties in the observations.

Finally, in 1994, came the smoking gun—Hubble Space Telescope observations of the core of the galaxy M87 that left no way out for anyone except the most intransigent black hole doubters. Hubble's evidence came from its view of a rotating disk of gas around the center of M87, a spiral galaxy about 50 million light-years from Earth, in the constellation Virgo. The high speed of the swirling disk indicated that the M87 core contained a mass of nearly 3 billion suns— so much that it must almost certainly be a black hole. Other examples followed, providing even stronger evidence. By the end of the century, it was clear that Einstein's universe was inhabited by many black holes.

Meanwhile, black holes also inhabited the minds of many theoretical physicists struggling to understand the nature of space, time, and gravity. During the 1990s, for example, superstring theorists began to take black holes very seriously. For it turned out, much to everybody's surprise, that black holes concealed clues to the mystery of the extra dimensions of space that superstring theory required. Black holes, it seemed, might merely be superstrings in disguise.

NEW DIMENSIONS

Superstring theory contained two distinct possible prediscoveries: the strings themselves, and the existence of extra dimensions of space. At first, all the excitement was about the strings; the extra dimensions were an embarrassment. But by the mid-1990s more physicists began to focus on the extra dimensions and what to do about them.

Of course, curiosity about extra dimensions was nothing new. The idea had been clearly prediscovered in literature, in the 1880s, by Abbott in *Flatland*. Other writers had discussed extra dimensions, and occasionally a scientist would speculate on the possibility. In 1893, the astronomer Simon Newcomb gave a talk in which he described the implications of a "fourth dimension" of space. "Add a fourth dimension to space, and there is room for an indefinite number of Universes, all alongside each other," he observed—which is just the sort of things that many physicists are saying today.[15]

Confusingly, Einstein's special relativity of 1905 also introduced a "new" dimension: time. Hermann Minkowski made this idea explicit in 1908, referring to the merger of time with space into spacetime, a continuum with four dimensions—three of space plus the one of time. Time therefore became known as the fourth dimension.

Time merely played the role of a fourth coordinate for describing a location. It was not at all the new dimension that Abbott had envisioned in *Flatland*. He was talking about a new dimension of space, just like the other three dimensions of space. But because of relativity's popularity, the idea of a fourth spatial dimension has to go by the name of *fifth dimension*.

In 1912, such a fifth dimension was proposed by the Finnish physicist Gunnar Nordström, as part of trying to incorporate electromagnetism into a theory of gravity. But it seems that Nordström viewed his extra dimension as just a mathematical trick, and in any event, he later abandoned his theory of gravity in favor of Einstein's general theory of relativity.[16] Physicists today generally trace the genealogy of extra dimensions in the modern sense back to Theodor Kaluza, a mathematician-physicist who was born in Ratibor, Germany (now Racibórz, Poland) in 1885. While he was struggling to survive as a young teacher at the University of Königsberg, Kaluza noted the similarities between Einstein's math for gravity and Maxwell's for electromagnetism. Perhaps the two theories might just represent a special case of an underlying unified mathematics, Kaluza speculated.

"If one consider this as a possibility," he wrote, "one is led almost inevitably to an initially unattractive conclusion"—namely, that such a view could be maintained "only by introducing the rather strange idea of a fifth space-time dimension."[17] Of course, "our previous physical experience contains hardly any hint of the existence of an extra dimension," he pointed out. On the other hand, nothing in our experience prohibits the existence of such an extra dimension, either—provided that any changes measurable in known physical quantities are restricted to the four ordinary dimensions. Any changes with respect to the fifth dimension could be either very small or zero, and so would not be noticed, Kaluza argued

In 1919 Kaluza sent a paper describing these ideas to Einstein, who apparently mulled it over for a while and then sent it off, in 1921, to be published. Kaluza's paper appeared in print shortly thereafter, before the development of quantum mechanics, and so it was strictly a classical approach. But by the mid-1920s, quantum fever had infected most of Europe's leading physicists, including Oskar Klein, a Swede who had studied under Niels Bohr in Copenhagen.

In 1923, Klein moved to the University of Michigan, where he worked out an approach to unifying gravity and electromagnetism with a fifth dimension. Klein was very excited about it until he returned to Europe and found out, from Pauli, about Kaluza's paper. Nevertheless Klein had gone a little farther than Kaluza, in particular realizing that the fifth dimension could escape detection by being very, very small. In fact, he calculated that a fifth dimension could be curled into a circle with a circumference of about 10^{-30} centimeters.

And Klein, unlike Kaluza, understood quantum physics and developed the idea in a quantum context. "Although the introduction of a fifth dimension in our physical considerations may seem rather strange at first sight," Klein wrote, quantum physics argued that an ordinary spacetime description of events on the atomic level was not possible, anyway. "The possibility of representing these phenomena by a system of five-dimensional field equations cannot be rejected a priori," Klein pointed out.[18]

He even suggested that a fifth dimension could shed light on one of the deepest quantum mysteries, the duality between waves and particles. By 1927 it was clear that particles sometimes acted like waves, and waves sometimes acted like particles. It could be, Klein said, that waves waving in five dimensions could produce what appeared to be particles in four dimensions. Klein's speculation sounds very close to the way string theorists today have begun to explain a new, richer concept of duality—a point we will get to soon.

Although the Kaluza-Klein five-dimensional approaches attracted some attention—from Pauli, for instance, and even from Einstein—they were largely forgotten. But decades later, the extra dimensions returned in a new package, tied up with superstrings.

THE MAGICAL MYSTERY THEORY

The first superstring revolution, in 1984, revived interest in the Kaluza-Klein approach, but the situation was much more complicated than it had been in the 1920s. Now one extra dimension was not enough. Superstring theory demanded at least nine dimensions of space—six extra ones beyond the usual three. Following Klein, superstring experts all surmised that the extra dimensions had to be extremely small. Obviously, any extra dimensions must be small, or so the reasoning went, because otherwise we'd have seen them by now, or people would fall into them and disappear. (Nobody discussed that possibility, though, for fear of encouraging wild ideas about the Bermuda Triangle.) And even if for some reason you couldn't fall into them, extra dimensions of space would alter physics in measurable ways. Space with more than three dimensions would affect the law of gravity, for example.[19] So the extra dimensions had to be "compactified"—rolled up into tiny little balls far too small to be seen.

To most of the string physicists, a few extra dimensions were no big deal, as long as the right math was available to describe them. All you needed to do was to find the math describing how six extra dimensions could curl up—in other words, how to describe the shape of space, or topology, that would contain six curled-up dimensions. Unfortunately, it seemed that thousands of different possible shapes existed in the equations.

With that realization, superstring theorists faced another serious problem. Remember, there seemed to be five different versions of string theory, all suitable for describing nature. Now, it seemed, even if you could decide which of the five versions of string theory actually did describe nature, you were left with thousands of variations of that theory, each with a different shape for space. "It was not very nice to have 10,000 unified theories," said Andy Strominger, a theoretical physicist now at Harvard, then at the University of California, Santa Barbara. "It would be nicer to have just one unified theory."[20]

But then, in mid-1995, Strominger and colleagues produced a phenomenal insight into the problem. Space, it turns out, could be constructed so that all the possibilities can exist. Space of one shape can transform itself smoothly into any of the other shapes, so you don't have to choose. Any one of them is just as good as another. A paper by Strominger, Brian Greene (then at Cornell), and David Morrison (at Duke) showed how the different possible shapes of the six extra superstring dimensions are just multiple ways of folding up the same underlying space, kind of like different knots in the same necktie. So perhaps there is only one way to represent space in superstring theory after all, and if there's only one way to represent space, it's a good bet that that space must be the one the universe is made of.

This realization came from insight into the superstring-black hole connection I mentioned earlier. At a fundamental level, described by quantum theory, a black hole is just like a basic particle of matter, the way water is inherently the same thing as ice. As Stephen Hawking pointed out in the 1970s, black holes "leak"; quantum processes allow radiation to get away, and in the process the black hole itself slowly evaporates. Ultimately a black hole could shrink into a tiny hole with about the mass of a bacterium. Under certain conditions, these tiny black holes can become entirely massless, in the process transforming themselves into superstrings.

Turning a black hole into a string might seem like the astrophysical equivalent of changing an elephant into an ant. Actually, though, it's more like turning a coffee cup into a saucer. Mathematicians describe such space-shape transformations as changes in topology. For a mathematician, changing a doughnut into a coffee cup is no problem (it just takes a little stretching and twisting), but there is no way to change a saucer into a coffee cup (without cutting and pasting—a violation of topological rules). But in the superstring view of space, such clever magic tricks become possible. At the same time that space can change from one shape to another, black holes can change into superstrings.

"Usually we think that no matter how much you stretch a coffee cup, you can't smoothly turn it into a saucer," said Strominger. "String theory does know how to smoothly turn a coffee cup into a saucer."

Still, transforming black holes into strings did not solve all of string theory's problems. There remained the five different versions of the theory. But that problem yielded in a similar way. The five versions of string theory were just different variations of one underlying theory—an actor appearing in different disguises, like Tony Randall in *Seven Faces of Dr. Lao*. The mysterious underlying theory was named M theory by Edward Witten, who proposed the idea. In a brief account published in *Nature* in 1996, Witten said the M could stand for magic, mystery, or membrane, according to taste.

Why membrane? Because M theory introduced a new dimension to superstring theory, literally and figuratively. Instead of 9 dimensions of space, in the old superstring theories, M theory required 10. And instead of all the objects being one-dimensional strings, higher-dimensional objects, called membranes, were permitted as well. Instead of rubber bands, the world perhaps was made of soap bubbles.

SUPERBRANES

In the late 1980s, as string theory progress had stalled, I ran across a new idea that struck me as no less strange than superstrings. I wrote a column about it, carrying the headline "Supermembranes offer a new dimension." It described an idea championed by Michael Duff, a physicist then in England, which took the original idea of superstrings a step further, into an additional dimension.

Superstring theory described tiny one-dimensional entities that curl into loops and vibrate, like plucked rubber bands, in 9 dimensions of space. Why not build the universe from two-dimensional objects—or membranes—in 10 dimensions of space? In fact, Duff contended, the extra dimension could solve many problems, and "supermembranes" might work better than superstrings at explaining how the universe is made. For some things, Handi-Wrap works better than rubber bands.

After all, if you believe in superstrings, accepting supermembranes is not that much of a leap. You can't argue that one makes less sense than the other. For that matter, a particle with no dimensions doesn't make much sense either, and it gives wrong answers to a number of important calculations, which is one reason superstrings became popular in the first place.

Alas, while I thought supermembranes sounded like a great idea, nobody else seemed to. Except for Duff, of course, who moved to Texas A&M University[21] and continued to pursue supermembrane studies. I asked him there, at a meeting in 1990, how the supermembrane idea had been received. "Hard-nosed string theorists would scoff at membranes, because they're dedicated to the string and that's the end of the story," Duff said. But it might be, he suggested, that both strings and membranes will turn out to have some validity.

"My own feeling is that neither . . . is actually the final theory of everything," he said. "My suspicion is that they're both just new layers of the onion skin—that the ultimate theory may embrace both of them but be more than either of them. But that's just a guess."[22]

Call Duff a good guesser. That's exactly what the situation seems to be today. When Witten introduced M theory, supermembranes (now just called "branes" for short) became a key part of the picture. But the picture itself became a little more confusing. Not only could you have both branes and strings, but branes could turn into strings. And any of the five versions of string theory could transform themselves into another version—just as one way of folding up the extra dimensions of space could switch into another. In other words, different versions of string theory were really all just the same theory looked at in different ways. Understanding how it all works demanded the expansion of the old notion of duality.

DUALITY

Duality is one of the deepest, and most difficult, concepts in modern physics. But it seems to hold the key to making sense out of the bewildering ideas of strings and membranes and extra dimensions.

At first glance, it isn't hard to grasp, though. Duality refers to a special kind of symmetry, in which two different descriptions of something turn out to be equivalent. Think of electrons, which have dual personalities—behaving like a wave in some experiments, like a particle in others, the way some people behave differently at work from the way they act at parties.

In a similar way, lots of things in nature appear to be different when viewed from different viewpoints. The back of a house may look nothing like the front of a house, yet it's the same house. (Even my cat can figure this out.)

But now for something a little more mysterious. Here's a reader participation quiz. Look at the two drawings, of a triangle and a circle (Figures 1 and 2). Explain how they are actually two drawings of exactly the same object.

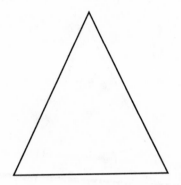

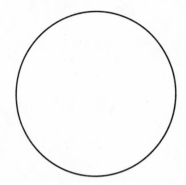

FIGURE 1 Triangle   FIGURE 2 Circle

How could that be? I haven't tested this one yet on my cat, but I'll bet many people could figure it out. It just requires a little creative thinking, the sort that superstring theorists employ. OK, I'll tell you the secret. You have to imagine an extra dimension. As long as you restrict your thinking to the two dimensions of the page, you'll never get the triangle to look like a circle. But if you add a dimension with your mind's eye, you can see how the circle and triangle are two views of a cone (Figure 3). In that way the two views are dual. In a very similar way, two theories that seem very different can turn out to be different views of the same theory. You just need extra dimensions to see how that can be.[23]

In M theory, of course, dualities get a lot more complicated. They involve adding dimensions in some cases, subtracting in others. They depend on what energy realms you are exploring—one theory describing phenomena at high energies turns out to be the same as another theory describing nature at low energies. A theory with membranes in 10 space dimensions can turn out to be the same as a theory describing strings in 9 space dimensions. In one way or another, each superstring theory is connected to another by a duality.

For example, in one version of superstring theory (Heterotic-E),

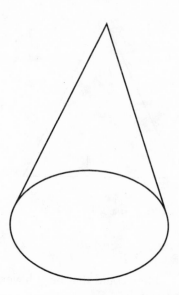

FIGURE 3 Cone

the math describes a one-dimensional string vibrating in nine dimensions of space. But that math works only if the string's "coupling" strength is low. (Coupling strength refers to how easy or hard it is to split a string into two new strings—a measure of how strongly or weakly strings interact. A strongly interacting string—with a high coupling constant—splits more easily. The lower the coupling strength, the harder it is to split the string.) Nobody knows whether the coupling strength in reality is high or low. If the coupling strength is high, the Heterotic-E superstring looks more like a supermembrane, requiring another dimension of space to vibrate in. The low-coupling version of Heterotic-E theory, in 9 spatial dimensions, is dual to another theory, in 10 spatial dimensions.

This example captures an essential aspect of duality. The two theories look very much different—so different that they are formulated in a different number of dimensions. Yet at some deeper level, the theories are the same—they describe the same physics. In principle, you could use either theory to describe the Heterotic-E string. But usually with dualities, one theory is a lot easier to use than the other. It's like a house with the front door locked; maybe you can get in through the back door.

In any case, the dualities connecting the various superstring theories suggest that they are all the offspring of some grander, bigger, all-encompassing umbrella theory—M theory. And for that reason, John Schwarz of Caltech likes to say that M should stand for *Mother*, as in Mother of all Theories.

EXTRA DIMENSIONS GET BIG

Working out the details of dualities, and other features of M theory, produced even more surprises. For one thing, it turned out that there should be a special kind of supermembrane, called a D-brane (named for Peter Gustav Lejeune Dirichlet, a nineteenth-century mathematician). Key insights into the importance of D-branes came from work in the 1990s by Joseph Polchinski of the University of California, Santa Barbara. D-branes turn out to be needed to accommodate non-loop superstrings. A string that did not form a loop had to have a place to attach its two loose ends. D-branes provided just such a surface.

An even bigger surprise emerged from work by Witten and Petr Hořava, who found an interesting approach to life in 11 dimensions (as M theory seemed to require). Perhaps the 11th dimension is a space separating two 10-dimensional "walls"—or branes. Standard Model forces and particles (including all of us) might live on one of those 10-dimensional boundaries. The other boundary would be a "hidden" world, not accessible to us.

But why would we be stuck on one brane? Maybe because it's a D-brane, and all the particles we are made of are strings that must remain connected to the D-brane surface. (In this respect, the D-brane becomes something like a black hole, from which nothing can escape. And in fact, work by Strominger and Cumrun Vafa at Harvard has shown a deep relationship between D-branes and black holes—in a sense, D-branes are like the spacetime bricks from which black holes are built.)

All this was very interesting mathematically, if somewhat lacking in relevance to daily life. But other strange ideas were floating around that eventually came together to make quite a splash. In 1990, for example, Ignatios Antoniadis at CERN suggested that superstring theory could accommodate a relatively large extra dimension. And in 1996, Joe Lykken at Fermilab pointed out that superstrings didn't necessarily have to be as supertiny as everybody thought. In fact, the influence of their existence might be noticeable at energies not too far out of the reach of current state-of-the-art atom smashers.

So it was in the air that something interesting might be going on in superstring theory's extra dimensions. And then, in 1998, a new line of inquiry popped into the extra dimension picture: the extra dimensions might be BIG!

A SILLY MILLIMETER BIGGER

Previous efforts to envision big extra dimensions had run into a big problem: gravity. Gravity is, after all, just the geometry of spacetime itself. So the strength of gravity would be affected by the presence of extra dimensions—it was a simple calculation. Gravity varies as the inverse square of the distance between two masses precisely because there are three dimensions of space. Three dimensions would dilute the strength of the force by exactly the amount predicted by the inverse square law. More dimensions would dilute gravity more than that. So any extra dimensions must be too small to make a big enough difference to measure.

Savas Dimopoulos, however, knew something that most other people working in the field didn't. A few years earlier, he had studied phenomena at distances shorter than a millimeter and found that gravity's strength had never been directly measured on those scales. It was thinkable, therefore, that gravity did *not* obey the inverse square law at short distances.

"It would be a disaster to change electromagnetism, but it would not be a disaster to change gravity," he told me.[24] "I knew there was much more freedom than people realized. I didn't know what it would lead to."

After our discussion in Houston about supersymmetry, Dimopoulos invited me to give a talk at Stanford, in February 2001, about my book *The Bit and the Pendulum*. At dinner after my talk, I asked him about the origin of the large extra dimension idea. The work leading to that idea, he said, was motivated by the hierarchy problem: why the fundamental mass in subatomic physics was so high compared to other basic quantities. Another way of phrasing it is to ask at what energy-scale gravity must become involved in explaining other phenomena—or, most simply, why gravity is (at ordinary energies) so much weaker than other forces. After all, a small magnet can pull a paper clip upward by magnetic force, even though the gravity of the whole Earth is pulling the other way.

"The key liberating thought was . . . what if only gravity propagates in the extra dimensions?" Dimopoulos said. He discussed the idea with Nima Arkani-Hamed, and then the two of them flew to Paris for discussions with Gia Dvali in early 1998. Out of those discussions came a breakthrough paper, appearing on the Internet in March and outlining a world with an extra dimension possibly as big as a millimeter—the size of a small ant, about 1/25th of an inch across.

That's nowhere near big enough to hold a mansion, let alone the whole universe. But the visible universe is huge only in the familiar three dimensions of space. In additional dimensions, the universe would be extremely slim—the way a sheet of paper is big in two dimensions but thin in a third. In other words, the whole universe is just one big supermembrane, or brane, with three big dimensions (and so is called a three-brane). Its other extra dimensions are too thin to notice. So our whole three-brane universe can fit into an extra dimension, even one a mere millimeter wide.

Think of the ant on a page of a book. The page has, to all appearances, two dimensions—width and height. Both width and height extend for several inches. Yet rip the page out of the book, and it could fit nicely in a file space a millimeter across, because the paper has an extra dimension, thickness, too thin to be noticed by the ant. In a similar way, the whole universe can have an extra dimension, too thin to be noticed by its human inhabitants.

Combined with the notion of the universe as a big brane, big extra dimensions suddenly became a realistic possibility. Maybe, physicists began to realize, extra dimensions are not invisible because they're so small, but because they're literally invisible. In other words, light can't go there. We wouldn't be able to see them even if they were big. And we couldn't explore the extra dimensions because ordinary matter, like light, was not allowed to go there, either. In this new view, all ordinary forms of matter and energy must stick to the surface of the three-brane universe. You could think of light and matter as made of strings that need to be anchored to a surface—the space of our brane. So light, radio waves, magnetism, quarks, electrons, all operate only on the three-brane—that is, in the universe's three familiar dimensions of space.

Gravity, though, can propagate as loops of string that don't need a surface to stick to. So gravity, and only gravity, is allowed to explore the extra-dimensional space, known as the "bulk."

With this realization, the imagination can run wild. In the Witten-Hořava picture, our universe could be viewed as one of the branes on the boundary of that slightly bigger 11th dimension.[25] But if our three-brane universe is thin enough to fit into a millimeter-wide dimension, so could another one. In fact, countless other ultrathin three-brane universes could be tucked into an extra dimension like pages of paper stacked in a file folder. In the hidden dimensions, the visible universe's thickness would measure on the order of a 10-millionth of a billionth of a millimeter. So countless such universes could fit in the extra dimensions.[26]

Such parallel 3-D universes, or brane worlds, might contain unusual forms of matter, possibly forming stars, planets, and people. "The specific laws of physics would be different in each of these branes," Joe Lykken explained. "Their law of gravity would be the same as ours, but everything else would be different. . . . But maybe they could form galaxies and stars and planets." And all would be less than a millimeter away from our "home" brane.[27]

LISA AND SUSY

Coping with an extra dimension as big as a millimeter is mind-bending (not to mention space-bending) enough. But that's not all. Maybe the hidden dimensions are not merely a millimeter wide, but perhaps are infinitely large. Our brane world universe could turn out to be just a bubble of foam in an endless ocean, a tiny island in a vast cosmic sea. The true totality of creation would extend beyond human sensation and imagination. Or at least beyond the imagination of most humans.

But not Lisa Randall's. In 1999, she and her collaborator Raman Sundrum published papers describing the math for a possibly infinite fifth dimension. I heard her mention this idea at a talk she gave at Fermilab in the summer of 1999, but I didn't pay close enough attention to figure it all out. So a year later I asked her to explain. To me, it seemed that the idea of an infinitely large extra dimension had appeared out of nowhere.

"Well in a sense it did," she told me. "We actually discovered this accidentally. We really started off motivated in an entirely different direction."

Randall and Sundrum had been contemplating an old problem with supersymmetry—or SUSY. Almost everybody thinks SUSY is beautiful, but anybody in love with SUSY has to face the ugly truth— her symmetry is badly broken, destroyed by some changing conditions in the universe's past (remember Chapter 3). If SUSY were faithful to nature, the superpartner particles would be all around us, with exactly the same masses as their partners. Evidently, that's not the way it is—the superpartners must be much, much more massive. Somehow, sometime back in the distant past, nature's supersymmetry was broken. It would be nice to know exactly how that happened.

The trouble is, many of the ideas for breaking SUSY also imply other phenomena that are known not to happen. Randall and Sundrum found, though, that adding extra dimensions to the calculations help avoid some of those unwanted consequences of SUSY breaking.

"That was our initial project, and it actually works quite well," Randall said. But they encountered one situation where it didn't work so well, where the gravitons—particles carrying the force of gravity through the extra dimension—were behaving badly (that is, the math wasn't working out so well). Randall and Sundrum realized that this bad behavior could be related to the old question of why gravity was so weak compared to the other forces. And then it hit them—that problem could be resolved if gravity didn't have to be the same strength everywhere. If the strength of gravity could vary, you could have an extra dimension as big as you wanted.

"So I'd like to say that we had this great idea," Randall said, "but in fact what really happened was we discovered it by accident. But then we realized its implications."[28]

So now we have another way of explaining why gravity is so weak. It simply doesn't have the same strength everywhere in the infinite dimension. Gravitons should condense most densely around some brane. And not our brane. If our brane resides some distance from the prime gravity brane, gravity would naturally not be very strong here.

"Gravity is weak because the graviton likes to be somewhere else in the extra dimension," says Joe Lykken, who collaborated with Randall in elaborating that idea. "If we could move a little bit over in the extra dimension, gravity would look much stronger."[29]

The Randall-Sundrum scenario and the approach of Dimopoulos and his collaborators differ in many respects. But both have provided an intellectual impetus for exploring realms beyond the known universe in a way that science has never been able to do before. The possible prediscoveries latent in these explorations would be on a scale of significance equal to, or even greater than, the grandest insights of human history—the Earth is round, the Earth is not the center of the universe, the universe is expanding. Of course, confirming the reality of extra dimensions, or universes beyond our own, will not be easy. But there's hope.

SEARCHING FOR SPACE

Detecting the presence of extra dimensions and possible parallel brane worlds is not unthinkable. Since gravitons can fly freely through the extra-dimension space (the bulk), a nearby parallel brane might be detected through gravitational effects. Astronomers should notice objects in the visible cosmos behaving weirdly, as though under the influence of gravity from an unseen source. In fact, that's exactly what astronomers do see—that's why they say the universe is full of dark matter. Maybe the dark matter in the universe is really "transparent" matter, residing in nearby brane worlds and therefore invisible.

Another way to detect the extra dimensions would be by precisely measuring the strength of gravity at submillimeter distances. Since gravity weakens more rapidly with distance if there are more dimensions, it should appear to grow more strongly than Newton would have expected as you move objects closer and closer. Attempts at such experiments so far show no signs of a deviation down to about a quarter of a millimeter, suggesting that if large extra dimensions are real, there must be more than one of them. (The more the number of large extra dimensions, the smaller the distance where their effects could be noticed.)

More evidence for extra dimensions could come from particle accelerators, perhaps the Large Hadron Collider (LHC) now under construction at CERN. The LHC could create particles with enough energy to escape from the brane and enter the bulk.

"You could actually deform your brane and produce particles that move off into the extra dimensions," Lykken explained. Such escaped particles would reveal their departure through "missing energy" after all the other fragments in a particle collision had been accounted for.

Hidden dimensions also imply the exotic possibility that the CERN atom smasher could create tiny black holes. Such mini-black holes would probably go poof in an instant, producing a burst of radiation that scientists could immediately recognize as a black hole's signature.

"You'd say, 'Aha! I've made a black hole,'" Lykken commented.

INTO THE UNKNOWN

Of course, all these ideas may turn out to be completely wrong. Rocky Kolb, while interested in the extra-dimension developments, remains skeptical. "Land speculation in the extra dimensions is not warranted at this time," he jokes.[30]

Lisa Randall is more confident. "At this point, I do think that the extra dimensions really are there," she says.[31]

Lykken also believes that extra dimensions are a real possibility. And he emphasizes that the question of higher dimensions and parallel universes is no longer just fun fiction, but has truly become a part of the scientific enterprise.

"You can go out and do an experiment in our lifetimes that will test whether these things are really there," Lykken said. "So it's not just fantasy, it's experimental science."

In any case, Lykken stressed to me that the point isn't that these ideas are right, but that they illustrate how profoundly little scientists really know about the ultimate shape of reality. "We know almost nothing about what the universe might be like in extra dimensions," he said. "We don't know how many extra dimensions there are, how big they are, what kinds of stuff live there. . . . We're going to find in the next century that there are all kinds of just amazingly weird things, and that we have not yet begun to make all of the discoveries that we're going to make, in physics and in all other fields. Physicists have been lulled into a sense of self-satisfied security that we know almost everything. And undoubtedly that's wrong. We don't know almost everything. In fact, we may know almost nothing."[32]

GHOSTS

From Riemann and the Geometry of Space
to the Shape of the Universe

In a finite universe . . . light which is received from opposite directions
may in fact have originated from the same location and simply took
different paths around the finite cosmos.

—Janna Levin and Imogen Heard
  "Topological Pattern Formation"

The skies might even contain facsimiles of the Earth at some earlier era.

—Jean-Pierre Luminet, Glenn Starkman, and Jeffrey Weeks
  "Is Space Infinite?"

Most scientists don't believe in ghosts. But some astronomers do.

Such ghosts have nothing to do with Christmases past or yet to come, but rather with the past and future of the universe. These ghosts are galaxies, or rather, images of galaxies, the huge groups of stars that show up in telescopes as spiral pinwheels or elliptical blobs of light. It may be that some of the galaxies in the sky are illusory, mere copies of galaxies already identified, showing up in another direction like the multiple images created by fun-house mirrors. Such ghostly images could appear if space throughout the universe is shaped rather strangely, allowing light from a distant galaxy to arrive at Earth by more than one route.

Strange as it sounds, this idea makes perfect sense to mathematicians who specialize in the branch of mathematics known as topology. Topology is concerned with the shape of space. Its formulas describe how the points that make up a surface, or any space, are connected.

Cosmologists have generally persuaded themselves that the topology of the real space of the universe must be simple. It's a lot easier to study and describe a universe with a plane-Jane space and no fun-house-mirror distortions. But there's no real evidence that space is so simple. On large scales, space's topology could be what the experts call "nontrivial," twisted around in such a way that you could see one and the same galaxy at different places in the sky.

All the triumphs of the big bang theory of cosmology have revealed nothing about the global shape of space. Data from early twenty-first century satellites will be needed to determine whether all galactic images are real or some are ghosts, and to tell astronomers how big the universe is—whether it is finite or infinite. If, in fact, the results show that the topology of space is nontrivial, it would be the most astounding restructuring of human conceptions of space since the prediscovery of non-Euclidean geometry.

NON-EUCLIDEAN GEOMETRY

In the nineteenth century, several mathematicians noticed that the standard geometry, handed down from the ancient Greeks in the form of books compiled by someone known as Euclid, might not be the only way that geometry could be done. Until then, for more than two millennia, Euclid's conception of space was the only one that anybody thought made any sense. His geometry was built with pure logic, founded on some simple definitions and assumptions that seemed unassailable.

Nobody knows much about Euclid, so it's hard to say whether he regarded his geometry merely as a logically consistent system or a true description of reality. But followers of Euclid clearly believed that his geometry described the physical world—it told how real space is shaped. Euclid's space is shaped, for example, such that the sum of the three angles in any triangle should amount to exactly 180 degrees.

In fact, many believed that Euclid's was the only geometry that the human mind was capable of knowing. Immanuel Kant, the influential eighteenth-century philosopher, taught that Euclidean geometry was ordained to be true by the very structure of the human intellect; space could not even be conceived to be otherwise. Therefore, Kant concluded, Euclidean geometry was an example of synthetic a priori knowledge—it was a truth about the world that could be known to be true without doing any experiments.

Kant, of course, was all wet, but many people believed him anyway.[1] Carl Friedrich Gauss, however, saw through Kant's overconfidence.

Gauss, the greatest mathematician since Newton, realized that geometry could, logically, be construed in a way different from Euclid's. Unfortunately Gauss was a perfectionist, who was therefore reticent about publishing his new ideas, especially those that were speculative. But his correspondence reveals that he had given serious thought to the subject of non-Euclidean geometry.

Born in 1777, Gauss first expressed Euclidean doubts as a teenager. In 1799 he confided some of those doubts to the mathematician Farkas Bolyai. "The path I have chosen," Gauss wrote of his studies, "seems rather to compel me to doubt the truth of geometry itself."[2]

And by that, of course, he meant Euclidean geometry, the only geometry anybody knew about back then. But that was soon to change.

Like many other mathematicians, Gauss was troubled by one central aspect of Euclid's geometry, the famous parallel postulate. Much of what Euclid proved depended on his assumption about parallel lines—namely, that if you specified a point not on a line, only one line passing through that point could be parallel to the first line. (There are various other ways of saying the same thing, but that's the idea.)

Before Gauss, few if any mathematicians doubted that assumption. But many believed it to be less than self-evident. Or they believed it to be so necessarily true that it ought to be possible to deduce it from Euclid's other assumptions. All such efforts to provide such a proof failed, however, though occasionally some geometer would erroneously claim success.

Even Gauss worked on such a proof, producing one that he thought would look good "to most people," but not to him. "In my eyes it proves as good as nothing," Gauss wrote to his friend Bolyai.[3]

Over the years, Gauss gave up trying to prove Euclid right about parallels and instead found ways to do geometry differently. Without the parallel postulate, other familiar features of Euclid's geometry were no longer necessarily true. The angles of a triangle, for example, did not have to add up to 180 degrees. "The assumption that the angle sum [of a triangle] is less than 180 degrees leads to a curious geometry, quite different from ours [Euclidean]," Gauss wrote in 1824 to another friend, Franz Adolph Taurinus. "The theorems of this geometry . . . contain nothing at all impossible."[4]

But since Gauss didn't like to publish anything that hadn't been worked out thoroughly enough to guarantee immunity from criticism, he left his non-Euclidean musing to letters. The first to propose a serious non-Euclidean geometry publicly was the Russian mathematician Nikolai Lobachevsky.

Lobachevsky, born in 1792, was a pretty bright kid. He started school young and at age 14 he entered the new university of Kazan. He excelled at math, earning academic honors, but before graduating he almost got kicked out of school for playing too many practical jokes. But by 1811, at age 18, he received his master's degree and then stayed at Kazan for 40 years, ultimately becoming the rector (the head guy) at the school that almost expelled him.

Besides teaching math, and physics, and astronomy, Lobachevsky filled other odd jobs at Kazan, such as organizing the library and the museum. Nevertheless he found time to pursue some original mathematical thinking, and by 1826 he'd produced, in the words of E. T. Bell, "one of the great masterpieces of all mathematics and a landmark in human thought."[5]

Lobachevsky had worked out the basics of a new geometry, based on the proposition that you could draw at least two parallel lines through a point not on the first line. He first delivered his ideas in an 1826 lecture; they were published (obscurely) a few years later. He called his creation "imaginary geometry" and claimed that it was superior to Euclid's.

"In [Euclidean] geometry I find certain imperfections which I hold to be the reason why this science . . . can as yet make no advance from that state in which it came to us from Euclid," Lobachevsky wrote later in a book explaining his system.[6]

He was not the only mathematician to explore the non-Euclidean realm at that time. Equal credit for discovering the new geometry sometimes goes to the Hungarian Janos Bolyai, son of Gauss's friend Farkas. Bolyai's approach was similar to Lobachevsky's, showing that Euclid's geometry was not the only possible description of the world. Or if it was, maybe there was another world. "Out of nothing," the younger Bolyai wrote, "I have created a strange new universe."[7]

He published his version in 1832, as an appendix to his father's math textbook. It was not a good place to attract a lot of attention, so Bolyai's work went unnoticed for years—as did, for the most part,

Lobachevsky's. Even those who knew about the new geometries were generally not impressed. For most of the nineteenth century almost everybody still believed Euclid. The "non-Euclidean" geometries were considered to be either utter nonsense or at best interesting exercises of the mind that had nothing to do with the real world. (Lobachevsky didn't help by calling his invention "imaginary" geometry.)

Bernhard Riemann, however, clearly understood that the real world might be different from the way Euclid envisioned it.

RIEMANN

Georg Friedrich Bernhard Riemann was another one of those tragic figures, a brilliant mind who died before his time. But he did manage in his 40 years to provide an abundance of intellectual fruit for later mathematicians—and physicists—to harvest. Riemann's math made Einstein's success of general relativity possible. And Riemann provided foundations for all sorts of further mathematical and cosmological inquiry.

In a way, it may have been beneficial for future scientists that Riemann died when he did, for had he lived he might have left future generations little to do. "It is quite possible," wrote E. T. Bell, "that had he been granted 20 or 30 more years of life he would have become the Newton or Einstein of the nineteenth century."[8]

Riemann was born in Breselenz, Bavaria, in 1826, son of a Lutheran minister, second of six children in a happy but very poor family. Shy and sickly, the young Bernhard was for years taught at home by his father, later moving in with his grandmother to attend a school in Hannover.

As a child he took a special interest in history, but by young adulthood his mathematical skills emerged. He astounded one teacher by mastering a 900-page book on number theory, by Legendre, in six days.

In 1846 Riemann entered the university at Göttingen—to study theology (his father's idea). Soon, however, he switched to math, and he attended lectures by Gauss himself. But on the whole, times were bad in Göttingen in those days, so Riemann went to Berlin, where there was no Gauss, but there were several other outstanding mathematicians, including Dirichlet and Jacobi. After two years at Berlin, Riemann returned to Göttingen to finish preparations for his Ph.D. He defended his dissertation in 1851. Gauss praised it effusively.

Then came the decisive step for the future of science and mathematics. Getting a Ph.D. in those days was nice, but good for nothing. Earning the right to lecture at the university required a candidate to prepare advanced work on a special area of knowledge and then to deliver a major "habilitation" lecture on a topic chosen by the university's review committee.

By the end of 1853, Riemann had completed his advanced report; he then offered the review panel three topics for his lecture—two on electricity (Riemann's preference) and one on geometry. Gauss swayed the committee to select geometry, a fortunate twist in history for the future of science. On June 10, 1854, Riemann presented one of the greatest lectures in the history of mathematics: "On the Hypotheses that Lie at the Foundations of Geometry."

It is hard to convey the richness of Riemann's lecture, its depth of insight, and its freedom from age-old prejudices. Somehow Riemann saw through the blindfold of tradition that had retarded human intellect for millennia. Geometry, Riemann noted, presupposes the concept of space and requires certain basic notions about points and lines to be taken as given, in advance of any application of logic. "The relations of these presuppositions," he declared, "is left in the dark. . . . From Euclid to Legendre, to name the most renowned of modern writers on geometry, this darkness has been lifted neither by the mathematicians nor by the philosophers who have labored upon it."[9] Riemann proceeded to lift the darkness.

Euclidean geometry seemed consistent with experience, of course, but perhaps that was only because measurements of it had never been precise enough to detect any deviations, he said. After all, Euclid's system had never been tested in the realm of the "immeasurably large" or "immeasurably small."

The realm of the small seemed particularly important, Riemann noted. "Knowledge of the causal connection of phenomena is based essentially upon the precision with which we follow them down into the infinitely small," he said. "One pursues phenomena into the spatially small in order to perceive causal connections, just as far as the microscope permits."[10]

But at the shortest distances, common human experience no longer is a sound guide. Euclidean geometry's natural fit with reality may be limited to realms of ordinary sizes. "It is entirely conceivable that in the indefinitely small spatial relations of size are not in accord with the postulates of [Euclidean] geometry," Riemann declared. The geometry of space itself may not be simple and the same everywhere, he noted, but could depend on "colligating forces that operate upon it." From the modern point of view, it's easy to see that Riemann's remarks foreshadowed Einstein's general relativity, in which matter and energy do in fact affect the local geometry of space.

In any event, deciding what space is "really" like cannot be properly done by logic alone—or by relying only on experience to date. Further observations of nature must be made, in realms beyond those previously accessible. But logical deductions from general notions, Riemann said, are valuable in making sure that physical investigations are not "hindered by too restricted conceptions, and that progress in perceiving the connection of things shall not be obstructed by the prejudices of tradition."[11] Whereupon, Riemann noted, the path of his inquiry would lead to physics, beyond the scope of his lecture on geometry.

It seems that among those who heard the lecture, only Gauss was smart enough to appreciate it. But later generations have profited enormously from Riemann's revelations in that lecture and their subsequent development.

His version of non-Euclidean geometry is the most famous of Riemann's contributions. It differed dramatically from that of Lobachevsky and Bolyai. Their "space" was curved inward, like a saddle, so that a triangle's angles would add up to less than 180 degrees. Riemann's was curved oppositely, like the surface of a sphere, so that triangles possessed angles adding up to greater than 180 degrees.

That is really not so mysterious—the same is true for triangles on the surface of a sphere. It's a property of curvature, and it's easy to see that a two-dimensional surface can be curved. But Riemann saw further, realizing that space itself could possess geometrical relations analogous to those of a curved surface. In fact, you do not need to restrict the math to three-dimensional space—you can describe curvature of space of any number of dimensions. Riemann developed the math for describing such multidimensional spaces, or manifolds. He opened up a whole new way for reasoning about space beyond the realm of ordinary appearances.

Riemann's approach naturally renewed the big question—does non-Euclidean geometry have anything to do with the "real" geometry of real space?—in a more sophisticated way. And that question, I think, illuminates the deeper issue concerning the relationship of mathematics to reality. Though Euclid's theorems are often portrayed as strictly logical structures, divorced from the physical world, his axioms and postulates had after all been based on what was "self-evident" from experience. But then along came Lobachevsky and Bolyai and Riemann, adopting a parallel postulate that is not self-evident at all, to produce a different description of space, with observable consequences. Rather than taking what was presented by the senses as the obvious truth, the mathematical minds of the non-Euclidean geometers inferred a different truth. And it turned out to be a truth not only "logically" true within its own system, but a truth about the physical world. And all this strikes me as a compelling case for the concept of prediscovery. The "logic" based on actual physical observation of the world produced the wrong geometry. The purely mathematical reasoning, exploring axioms *not* derived from experience, turned out to produce the prediscovery that space itself is curved.

Of course, Riemann did not live to see that vindication. It was half a century after Riemann's death before Einstein, in attempting to understand gravity's relationship to space, found that Riemann's math was just what he needed to make his theory work.[12] In fact, for a long time while working on general relativity Einstein was hopelessly stuck. It was only after his friend Marcel Grossmann alerted Einstein to the existence of Riemann's math that general relativity's pieces began to fall into place. Einstein's success in formulating general relativity vindicated Riemann's intuition that actual space might not be quite what Euclid had taught. "I have shown how Riemann's theory . . . can be utilized as a basis for a theory of the gravitational field," Einstein declared in 1915.[13]

In 1919, an eclipse offered the opportunity to test how the geometry of space depended on the presence of a massive object, in this case the sun. Light from a distant star chose the path that corresponded to Einstein's version of Riemann's geometry. Riemann, I therefore conclude, had prediscovered the true nature of the geometry of space. (If you like, you can give Einstein some of the credit, too.)

COFFEE AND DOUGHNUTS

Nowadays another grand question about space occupies the minds of freer thinkers who appreciate other aspects of Riemann's legacy, having to do with space's topology.

Different writers construe the terminology of topology in various ways. Some would say geometry includes both local features of space (designated by the term *metric*) and global features of all of space (described as space's *topology*.) But more often, I think, experts refer to local features of space as *geometry*, and global features as *topology*.

There is another distinction. Geometry is about measurement and quantity; topology is about position and place. Geometry describes the precise relationships of angles and distances. Topology describes more general relationships of the points in a space. If you draw a circle freehand, it will probably not be perfectly round. Precise measurements will show that it does not have the exact geometry of a true circle. But it does have the topology of a circle, a line of points closed in a loop.

We've already discussed how the geometry of space is warped by the presence of matter, making it curved.[14] And we've seen that the universe has an "average" geometry, determined by the amount of matter (and energy) that the universe contains. But whatever geometry the universe exhibits on average, local geometry can be changed by adding mass or taking it away.

Topology, on the other hand, is forever. It describes the intrinsic shape of space as a whole—such as whether space has holes in it.

The textbook example is space corresponding to the surface of a doughnut. A doughnut has a hole. So would the space corresponding to a coffee cup, with its hole in the handle. A doughnut made of tough and gooey dough could be deformed to resemble a coffee cup. But you could not reshape a ball into a coffee cup without cutting a hole in the ball somewhere. Doughnuts and balls thus have different topologies.

A shape with no holes is what mathematicians call "simply connected." Technically, the key issue is whether any closed curve on a surface can be shrunk down to a point. Draw a circle on the surface of a ball, for example, and you can see that you could make it smaller and smaller; ultimately it would look like a dot. But draw a loop on a doughnut through the hole and back around the outside. You can't keep shrinking that loop; the doughnut gets in the way.

It's a little trickier to visualize the topology of empty space, but the mathematical principles are the same. Space itself might be "shaped" in such a way to have very much the same properties as a surface with a hole.

Imagine, for example, a space constructed by stacking a bunch of cubes together to make a bigger cube. (Remember Rubik's cube? Think of something like that, without worrying about the colors.) Now imagine that everything inside one cube is identical to everything in the next cube over (and above and below). Suppose further that the middle of one cube contained the Milky Way galaxy (and therefore Earth, you, and this book). In the next cube over (and above and below) would sit another Milky Way, another Earth, another you—a few billion light-years away.

In a "repeating" set of cubes like this, space is connected in a nontrivial way. The right side of one cube corresponds precisely to the left side of the next cube. In principle, you could look out into the next cube over and see the back of your own head.

Only recently have cosmologists begun to consider the possibility of such nontrivial topology seriously. But the idea actually goes back to the prescient writings of Karl Schwarzschild, the German mathematician who first showed that Einstein's theory of general relativity implied the existence of black holes. A century ago, Schwarzschild commented that space could be put together in tricky ways. Imagine, he said, that astronomical observations deep into space revealed a curious repetitious pattern; as we looked farther and farther out, we would see more and more galaxies identical to the Milky Way. (Remember, at that time the Milky Way was the only galaxy known.) It would merely appear, Schwarzschild proposed, "that the infinite space can be partitioned into cubes each containing an exactly identical copy of the Milky Way. Would we really cling on to the assumption of infinitely many identical repetitions of the same world? . . . We would be much happier with the view that these repetitions are illusory, that in reality space has peculiar connection properties so that if we leave any one cube through a side, then we immediately reenter it through the opposite side."[15]

Mathematically, Schwarzschild made perfect sense. But nobody paid much attention. Why would anyone want to contemplate space shaped in such a strange way? A little over two decades later, though, the issue acquired some pertinence in connecting Einstein's general relativity theory to the description of the universe. Alexander Friedmann, in his papers establishing the possible expansion of the universe, warned that predicting its future required a knowledge of the topology of space. It would be a mistake, he said, to assume without evidence that space was shaped in the simplest imaginable way. But then Friedmann died young, physicists forgot his warning, and only today have they begun to realize that the shape of space remains undiscovered—and therefore that space could be full of ghosts.

FEAR OF INFINITY

Pursuing these illusory images today are a host of cosmological ghost-busters—clever thinkers like David Spergel of Princeton, Jean-Pierre Luminet of the Paris Observatory, Glenn Starkman of Case Western Reserve in Ohio, and Janna Levin of Cambridge University, in England. They and others have analyzed the possibility of ghost images in the universe and various methods for detecting them. Their motivation is not to perform a cosmic exorcism, but to avoid what many people regard as the unsavory philosophical consequences of an infinite universe.

Einstein himself shared this concern. His equations did not specify the shape of space for the universe as a whole but only indicated how space would be curved in the presence of matter and energy. You should be thoroughly familiar by now with the three possibilities: Einstein's equations allow the universe to be "closed" (curved like a ball), "open" (curved like a saddle), or "flat" (like a sheet of paper). A closed universe would be finite. A flat or open one would be infinite—*if* space's global topology is simple.

Einstein clearly preferred a closed, finite universe. That prejudice reflected the influence of Ernst Mach, the physicist-philosopher whose disdain for atoms inspired my tirade in Chapter 7 about the possibilities for detecting superstrings. To the general public, Mach's name is best known today for describing velocities greater than the speed of sound. Among physicists, though, he is better known for (besides his disbelief in atoms) a view of the universe known as Mach's principle. Boiled down to its essence, Mach's principle decrees that the mass of an object depends on all the other objects in the universe. More technically, Mach was talking about inertia, the tendency of an object to resist change in its state of motion, which is in fact a measure of its mass.

Einstein's general relativity was built on the principle that the mass as measured by inertia is precisely the same as the mass involved in gravity. So Einstein's theory incorporated a natural affinity with Mach's principle. But both suffered from a serious problem in an infinite universe—the mass of such a universe would also be infinite. And therefore so would the inertia of any object. For that reason Einstein preferred the form of his equations in which the universe was closed, and therefore finite. When mathematicians objected that nontrivial topologies would allow even an open universe to be finite, he argued that trivial topology should be preferred because of its simplicity.[16]

For several decades afterwards, nobody seemed to worry too

much about the topology problem. Textbooks and popular articles alike generally ignored the topological loopholes that complicated the cosmological story. But as the twentieth century came to an end, astronomical observations forced physicists to reexamine their assumptions. As the evidence began to suggest an open universe, the drawbacks of an infinite space drew more attention to themselves.

In particular, studies in the 1990s of light from distant stellar explosions indicated that the universe is now expanding faster than it used to be. That would seem to suggest that the universe will expand forever, and not collapse someday, as it would if the universe is closed and finite. That development inspired a new round of fear of infinity.

INFINITY'S CURSE

In addition to the Mach-Einstein objection, infinity posed other serious philosophical problems. I remember well a physics meeting in 1998, where several physicists discussed the implications of the accelerating expansion of the universe.

At the time, the evidence wasn't yet wholly convincing—at least not to me—and even now a few cosmologists raise some doubts about the conclusion that the universe will expand forever. Nevertheless, there was more reason than ever to believe that the universe might be infinite in extent. And that raised some disturbing issues.

"There's some very speculative and bothersome and almost philosophical problems with actually infinite universes, even though that infinity is somewhere over the horizon beyond what we can observe," said Edwin Turner, a Princeton astrophysicist.[17]

For instance, in a truly infinite universe, all possibilities become realities. An infinite universe would encompass an endless number of additional regions of space, equal in size to what astronomers can already see. But each region would contain a limited number of atoms, which could be arranged in a limited number of ways. With no limit on space, all the possible atom arrangements would recur over and over again. So all possible combinations of matter, and all sequences of activity, would happen somewhere out there. Every person, every event, would exist in multiple places. Every baseball game and presidential election would replay itself. "If the universe is really infinite," said David Spergel, "we're having this conversation an infinite number of times, right now." Or as Janna Levin puts it, "Somewhere else in the cosmos, you are there. In fact there are an infinite number of you littering space."[18]

I talked to Spergel again three years later, shortly before the launch of the MAP (for Microwave Anisotropy Probe) satellite—Spergel's best bet for finding out whether the topology of the universe is trivial or not.

"Geometry and topology are related but are not the same," he reminded me. "Geometry is local curvature, topology is the large-scale structure. . . . If the geometry is flat or negatively curved, then the topology can either be infinite or finite. You can either have a flat sheet of paper that goes on forever, or you can fold it on itself and connect it up."[19]

In fact, he said, if the universe is negatively curved—as the supernova evidence seemed to indicate—there are an infinite number of distinct ways to fold up space. So cosmologists who prefer a finite universe must figure out a way to show that the topology of space is nontrivial and how one of the many possible foldings would make the universe look the way it does.

Actually, Spergel insisted, nontrivial topology is not so unfamiliar. "People have a lot of experience with nontrivial topology from video games," he pointed out. A video character moving off the right side of the screen can instantly reemerge on the left, as though the right side of the screen's "space" were somehow connected to the left. Perhaps the space of the universe is connected in a similar way. Instead of a series of identical TV screens, the universe might consist of identical cubes, as Schwarzschild envisioned. More accurately, the universe would be one big cube with all its sides connected. So if you looked out into space, you would see what seemed to be identical cubes surrounding a "central" cube on all sides.

If so, powerful telescopes looking deep into space might reveal galaxies in apparently adjacent cubes, but they would really be galaxies in the same cube, just seen from a different direction. One of the "distant" galaxies could turn out to be a ghost image of the Milky Way itself. A telescope pointing away from the Milky Way's center could see it from the other side. (You could argue, of course, about

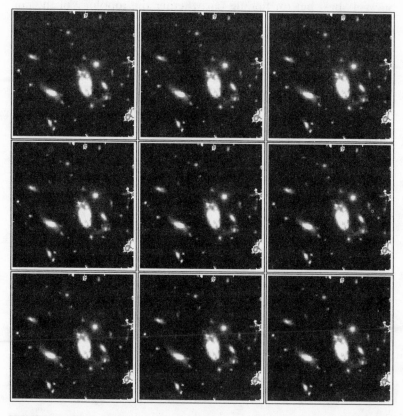

FIGURE 4   Strangely shaped space could create the illusion of multiple images of the same galaxies.

which image is the real galaxy and which is the ghost. Some experts say the closest image is the real thing and the others are ghosts, but on the other hand the closer image might be in an adjacent "cube." In that case it might make more sense to call the image within your own cube the real one. It's all just a matter of definition.)

But looking for ghosts in this straightforward way might not work too well. For one thing, if you see an object via light coming from different paths, the light will arrive at different times. A ghost of a galaxy will not look like what that galaxy looks like now but what it looked like a long time ago. So searching for identical galaxies at different points in the sky is probably a poor strategy. Some experts have suggested looking instead for similar clusters of galaxies, which would not have changed much over time. It might even work to look for groups of quasars, which would have changed in individual appearance but would have retained the same spatial relationships to one another.

MAP'S MAP

An ingenious scheme that might pay off sooner was proposed in the late 1990s by Spergel and colleagues Neil Cornish and Glenn Starkman. They plan to exploit data collected by the MAP satellite, launched in the summer of 2001. MAP was designed to record the temperature of the cosmic microwave background radiation, which has been streaming through space for 13 billion years, ever since the big-bang explosion cooled enough for atoms to form. For the most part, you'll recall from Chapter 5, the temperature of that radiation is the same everywhere; subtle deviations betray the spots where small clumps of matter appeared, forming "seeds" that grew into galaxies.

Finding nontrivial topology isn't its main purpose, but MAP's map of the microwave sky could be scanned for evidence of identical temperature deviations appearing in different locations. In fact, Spergel and colleagues calculated, a non-simple topology could reveal itself by imprinting precisely identical circles of temperature blips on the sky. (In other words, the blips in a circle around one point in the sky will be just the same as the blips around another circle at some other position in the sky. It will take some high-speed computing power to find such circular patterns.)

MAP could detect such circles only if the actual size of "one copy" of the universe (known as the "fundamental domain") is small enough. So far, Spergel told me, other kinds of searches have established that the fundamental domain—the actual cube—must be at least 2 billion light-years across. Otherwise ghosts would have already been identified. MAP will extend the search for nontrivial topology to the entire visible universe.

"If it's smaller than that, then we'll see it," Spergel said, and that would guarantee that the universe is finite. If MAP doesn't see the circles, the universe might still be finite, but the "cube" size would be bigger than the whole apparent universe (the "covering space"), about 60 billion light-years across. (That number seems too big for a universe that is only 13 billion or 14 billion years old. But Spergel points out that the 13 or 14 billion years is the time light has had to travel to us from the most distant sources. During that time, the universe has continued to expand. So if you estimate how big it "really" is "now," it would be more like 60 billion light-years across.)

So MAP may not be able to answer the topology question. "But it's worthwhile to spend three weeks of computer time to look," Spergel said, "given that you built that satellite."

Of course, even further complications are possible. Space could be connected in many different possible ways. The "cube" version is only one of the simplest. The mathematician Jeffrey Weeks has described a much more complicated shape—with 18 sides—that could serve as the fundamental domain of space. Other shapes with even more sizes are also possible.

To many scientists, such complicated multiconnected topologies seem to defy the Occam's razor approach of seeking simplicity in science. But Janna Levin and Imogen Heard argue that nontrivial topology is no more exotic than curved spacetime to begin with. And Jean-Pierre Luminet and Boudewijn Roukema point out that what counts as simple depends on your point of view. Modern attempts to describe the origin of the universe suggest that it might be simpler to make a universe with nontrivial topology than to make one with trivial topology. Nowadays the most popular way to "make" the universe involves quantum theory, which suggests there is some probability of a universe popping into being out of nothing. But the probability diminishes as the universe being created gets bigger. For an infinite universe, the odds are very slim.

"An infinite universe would have zero probability of coming into existence," Luminet, Starkman, and Weeks pointed out in a 1999 *Scientific American* article.[20] That suggests it would be a lot easier to make a finite universe. And current observations suggest that if our universe is finite, it must possess nontrivial topology.

Levin and John Barrow have proposed further benefits of nontrivial topology. It might explain, for instance, the highly structured arrangement of galaxies in the cosmos that seem to have grown from that smooth distribution of seeds in the cosmic microwave background radiation. Levin and Barrow invoke the murky math of quantum chaos to derive this conclusion, but their main point seems to be that structure might arise naturally along the shortest loops to get from one place to another in a nontrivial finite space.

"If . . . the universe is topologically finite," they write, "then light and matter can take chaotic paths around the compact geometry. Chaos may lead to ordered features in the distribution of matter throughout space." The distribution of galaxies, then, may be "providing a map of the shortest route around a finite cosmos."[21]

Levin and Heard suggest that nontrivial topology might be

connected with string theory, inflation, and the vacuum energy or cosmological constant. A nontrivial topology might explain why some of the dimensions in superstring theory are large while others are small, a process that might have been worked out during the epoch of inflation.

"The . . . magnitude of the vacuum energy depends on the topology," Levin and Heard write, "and it is conceivable that it selects three dimensions for expansion and three for contraction in a kind of inside/out inflation." Understanding topology, it seems, may solve many problems.[22]

If it turns out to be the case that the universe is finite, with a nontrivial topology, science would once again be presented with an astounding anticipation. Just as Riemann and his predecessors foresaw that space's geometry could be non-Euclidean, Schwarzschild and his successors forecast the possibility of nontrivial topology. Verification of their possible prediscovery may only be a matter of time.

THE TWO-TIMING
UNIVERSE

10

From Einstein and Slow Clocks to a Second Dimension of Time

One could entertain the idea that the real world requires more than one, or possibly all, of the theories permitted by the mathematics.

—M.P. Blencowe and Michael Duff
"Supermembranes and the Signature of Spacetime"

It's time once again to illuminate the frontiers of modern physics with movie trivia.

The film is *Frequency* (Dennis Quaid, Jim Caviezel, New Line Cinema, 2000). Through a ham radio mysteriously able to transmit 30 years into the past, Quaid learns from the 1999 version of his son (Caviezel) what will happen in the 1969 World Series. And the trivia question is: Who is the physicist appearing on a TV show running in the background during a scene early in the movie?

The answer is Columbia University string theorist Brian Greene, playing himself, but made up to look old and gray.

"String theory dramatically changes our understanding of space and time," Greene tells Dick Cavett, also playing himself. "For example, it turns out that string theory requires our universe to have ten or possibly even eleven dimensions. And the strange thing is, some physicists are even pursuing the idea that there might be more than one time dimension."

The even stranger thing is, that part of the movie wasn't fiction. Some physicists really *are* pursuing the idea of a second dimension of time.

"So in addition to time as we know it," Greene explains, "there may be a second time dimension where the universe evolves in some different manner."

That's about where the physics in *Frequency* ends, though. Greene didn't get enough screen time to explore the ramifications of a second time dimension. But the theme of the movie does resonate with a realization that often strikes scientists and nonscientists alike: there is something mysterious about time. And all the advances of the past century in physics, while clarifying so much else about reality, have mainly deepened time's mystery.

Just what the idea of a second time dimension means, nobody really knows. So as usual, physicists resort to concealing the mystery with a secret code. Remember, they don't know the details of the ultimate theory of everything, so they call it M theory, for "mother of all theories." Some theorists—Cumrun Vafa of Harvard University, for example—think M theory needs a companion. He calls the companion F theory, with the F standing for *father*.

F theory sounds like a way of giving fathers equal time, but in fact, time is not equal in the two theories. M theory, like all traditional theories, has just one dimension of time. F theory has two. F theory's second time dimension is no doubt what Brian Greene was alluding to in *Frequency*.

Many other physicists regard the idea of a second time dimension as too bizarre, even for the movies. But perhaps it is not all that much more bizarre than Einstein's prediscoveries about time, nearly a century ago.

EINSTEIN'S TIME

Nobody discerned more truths about the universe before they were discovered than Einstein. He foresaw the expansion of the universe and the vacuum energy or "cosmological constant." His math predicted black holes, even though he didn't believe in them. He realized that matter could be converted into energy and that space was curved. And he even foreshadowed the lyrics to the famous Chicago song of the 1970s: "Does anybody really know what time it is?"

Einstein not only asked the question, he answered it: No. Nobody knows what time it "really" is, because there is no one real time. If you try to order events in a time sequence—first to last—your list may not match that of someone who is moving in relation to you and those events.

Not only did Einstein show that different observers would place events in a different time order, he found that the rate of time itself would change for an object in motion—the faster it moved, the more slowly time would flow. It was one of the amazing consequences about the physical world that emerged from Einstein's 1905 special theory of relativity.

Special relativity, it seems to me, is underappreciated these days. General relativity is where the action is now, with its cutting-edge cosmological implications. All the attention to general relativity is certainly warranted. It produced many of the most profound prediscoveries of the past—black holes, and the expansion of the universe, for example—and retains many more potential prediscoveries up its sleeve, especially when you team it up with quantum mechanics.

But special relativity deserves a special place in the history of physics as well. It was, after all, the essential first step toward general relativity. And it was a singular example of how the greatest genius of the twentieth century achieved success in prediscovering natural phenomena whose existence no one else ever suspected.

In 1905 Albert Einstein was unknown in the world of physics. Unable to secure a teaching job after earning his degree in physics from the University of Zurich, he had gone to work in the patent office in Bern as a technical expert, third class. He didn't seem to be cut out for an academic career, anyway. His dislike for classroom discipline and his distrust of authority had caused his scholastic record to be less laudable than it might have been. While in elementary school, Einstein had shown little promise. When his father asked the school's headmaster what profession young Albert should choose, the reply was not encouraging: "It doesn't matter; he'll never make a success of anything."[1]

According to Einstein's own report, his distrust of authority originated at the age of 12 when he realized from his science readings that some Biblical stories he had been taught could not be true. At about the same time an uncle gave him an old geometry text, which young Albert devoured with considerable energy. Apparently this encounter with plane geometry was one of the few intellectually stimulating events of his youth; the only comparable experience was the gift of a pocket compass from his father when Albert was five.

Albert was no fan of the German school system, finding it suffocating and excessively rigorous. When his parents moved from Munich to Milan in 1894, he stayed behind to finish school. He soon quit, however, and followed his family to Italy just a few months before he was to have received his diploma.

After taking a few months off from education, he decided to try the Swiss system, applying in the fall of 1895 to the Federal Polytechnical Institute in Zurich. But he failed the entrance exam. So he spent a year in a Swiss high school, enabling him to get into the Zurich institute.

Einstein's nonconforming ways continued at Zurich. He attended few lectures, preferring instead to stay in his room reading the masters of nineteenth-century physics, such as Kirchhoff, Helmholtz, Maxwell, and Hertz. He did spend a lot of time in the lab, and he passed the final exam, thanks mainly to the helpful lecture notes of his friend Marcel Grossmann.

But Einstein graduated from Zurich disgusted with the educational system. "I found the consideration of any scientific problems distasteful to me for an entire year," he remarked. It was miraculous, he said, that "the modern methods of instruction have not yet entirely strangled the holy curiosity of inquiry."[2]

In Einstein's case at least, the system did not strangle his interest in physical science. Unfortunately, his preoccupation with physics entailed the neglect of mathematics. Einstein later commented that there existed too many branches of mathematics and no criteria by which to choose the most significant. "In physics," he declared, "I soon learned to scent out the paths that led to the depths."[3]

Upon graduation, Einstein sought employment in the laboratories of Heike Kamerlingh Onnes and Wilhelm Ostwald, but his inquiries went unanswered. For two years he struggled, making a little money by tutoring and substitute teaching. Then, in 1902, the father of a friend helped him get a job at the patent office in Bern. There, for seven years, he served as a technical expert—and there he produced some of science's greatest insights into nature.

Isolated from the academic world of physics, Einstein's mind was not cluttered by unnecessary knowledge or irrelevant distractions. His intuition was free to pursue its own perceptions of the physical systems he found intriguing. One of those involved a paradox he had first discerned at the age of 16, according to his own autobiographical testimony. "If I pursue a beam of light with the velocity c (velocity of light in a vacuum)," he reasoned, "I should observe such a beam of light as a spatially oscillatory electromagnetic field at rest."[4] Such a phenomenon, however, seemed not to exist; nothing like it had ever been observed. It corresponded to nothing in Maxwell's theory of electromagnetism, either. Einstein eventually concluded that there appeared to be no way to assert that a system is in a state of absolute rest. Motion, in other words, is relative.

Throughout his education at Zurich and during his first few years at the patent office, the ramifications of these realizations swirled in his mind. Finally, in 1905, a conversation with his friend Michele Besso suddenly crystallized the latent revolution. In a matter of weeks, Einstein prepared the paper that spelled out the implications of his relativity principle. He entitled it simply "On the Electrodynamics of Moving Bodies."

Einstein built his special theory on two postulates. The first postulate: The laws governing two "reference frames," in uniform motion with respect to each other, are the same. The second postulate: Every light ray moves through empty space with a fixed velocity c, independently of whether the ray is emitted by a body at rest or in motion. The first postulate was Einstein's statement of the principle of relativity. Nowadays it would be regarded as a symmetry principle: the laws of nature stay the same no matter what direction or how fast you are moving. (It's just that in the case of special relativity, motion has to be "uniform"—in a straight line with constant speed.)[5]

The second principle, Einstein declared, is contained in Maxwell's equations. It's part of the laws of nature that the speed of light stays the same for all observers, no matter how fast they are moving (in a straight line at constant speed). Einstein's great insight was that these two postulates are compatible. His famous paper of 1905 began to work out the implications of that compatibility.

One significant implication concerned the notion of simultaneity. Einstein pointed out that you could find no objective point of view for deciding whether two events separated in space occurred at precisely the same time. Whether two events were simultaneous or not depended on the motion of the observer making the judgment.

Strange new conclusions also emerged about a moving body's mass. As a body approached the speed of light, its mass would increase; if it could attain the speed of light, its mass would become infinite. Therefore, it seemed, it would not be possible to accelerate an object with any mass at all to a velocity equal to that of light. The speed of light became a cosmic speed limit.[6]

Another curious effect of rapid motion was a foreshortening of the moving object in the direction of its motion. An observer moving along with such an object would notice nothing unusual. But an observer at rest would see a rapidly moving object appear to scrunch up—a ball, for example, would appear to flatten itself into something like a vertically oriented pancake. The amount of this scrunching increases as the velocity of light is approached. The exact degree of shrinkage can be calculated by a formula previously described by the Irish physicist George Fitzgerald and the Dutch physicist Hendrik Lorentz. (The shortening of an object in the direction of its motion is therefore referred to as the Lorentz-Fitzgerald contraction.)

It might be cheating to call this foreshortening of objects in motion a prediscovery, however. Lorentz and Fitzgerald developed their math to try to explain the famous Michelson-Morley experiment of 1887. But Einstein's explanation for the contraction did anticipate effects not known back then—the shrinking effect has to be taken into account, for example, when analyzing the impact of fast-moving subatomic particles in accelerator experiments.

Special relativity nevertheless was rich with true prediscoveries. Most dramatic, perhaps, was Einstein's deduction that mass and energy are equivalent, a point he spelled out in a subsequent 1905 paper. But the prediscovery most pertinent to this chapter involved a deep realization about the nature of time. A body in rapid motion, Einstein showed, experiences a slowdown in time relative to a stationary observer. Newton's "absolute, true and mathematical time," flowing "equably without relation to anything external" would no longer be the time of physics.

SLOWING THE CLOCKS

Einstein's 1905 relativity paper showed that the same math describing the Lorentz-Fitzgerald contraction would describe the changing rate of time for objects in motion. It seems paradoxical, but Einstein mentioned it in his paper in an almost offhanded way, referring to it merely as a "peculiar consequence" of his postulates. He did not remark on how astounding it must have seemed to others who read his paper, but he illustrated the idea pretty clearly.

Consider, he wrote, two clocks, at points A and B, both at rest with respect to a coordinate system K. Make sure the clocks are synchronized and both keep good time. "If the clock at A is transported to B along the connecting line with the velocity v, then upon arrival of this clock at B the two clocks will no longer be running synchronously," Einstein wrote. "Instead the clock that has been transported from A to B will lag . . . behind the clock that has been in B from the outset."[7] The precise amount of the time lag could be calculated using the Lorentz-Fitzgerald formula, with time replacing length.

Einstein went on to point out that similar reasoning applied if two clocks started out at the same spot. If one flew off and then returned, it would lag behind its stay-at-home counterpart. There was no getting around this conclusion. If Einstein's postulates were true of nature (and they certainly seem to be), nature *must* play some pretty clever tricks with time. Except this trick was no illusion. It is not a case of a moving clock just turning its gears more slowly than one at rest, or of its hands encountering friction. A person traveling alongside a rapidly moving clock would notice nothing wrong with the clock. Time itself slows down for the clock, the person, and anything else traveling along at the same speed (with respect to a time-piece at rest). The implications became clearer when, a few years later, this clock paradox was personified into what has become known as the twin paradox.

In a 1911 lecture, Einstein spelled it out. "Whatever holds for the clock, which we introduced as a simple representation of all physical phenomena, holds also for closed physical systems of any other con-stitution. Were we, for example, to place a living organism in a box and make it perform the same to-and-fro motion as the clock . . . it would be possible to have this organism return to its original starting point after an arbitrarily long flight having undergone an arbitrarily small change, while identically constituted organisms that remained at rest at the point of origin have long since given way to new gen-erations."[8]

Therefore if one of a pair of identical twins takes off in a fast-flying rocket ship while the other twin remains homebound, the stay-at-home twin will age more rapidly, because time is slowed for the traveler.

People puzzled by this aspect of relativity still sometimes com-plain that the twin paradox cannot be the way it seems. If motion is relative, why can't the stay-at-home twin pretend to be the one rap-idly moving? If you restrict your analysis to special relativity, that question is hard to answer. Because the only way to test the question is for the flyboy twin to return to Earth, and to do so requires some maneuvering in space that breaks the special relativity rule about moving only in a straight line at a constant speed. In other words, the two twins do not have equivalent experiences, and the flying twin will indeed age less rapidly. (Actually, you can devise situations in which the moving twin could return younger without breaking the special relativity rules, too, but that gets a little more complicated.)[9]

In any event, the twin paradox (or as it is more properly called, the time dilation effect of special relativity) is to me a clear-cut example of prediscovery. A century ago Newton's absolute time seemed pretty self-explanatory to most people. The idea that the objective time of physics could slow down strikes me as utterly outside any actual physical experience or evidence. Yet Einstein's theory, with help from the Lorentz-Fitzgerald math, revealed this aspect of the real world in advance of its discovery.

In fact, it took quite a while for that actual discovery to take place. The first really solid evidence came in the early 1940s, based on measurements of subatomic particles called muons. Muons are created in the upper atmosphere by cosmic rays striking air atoms. But muons are unstable and decay very rapidly, within a microsecond or two on average. Nevertheless many muons make it to the ground, a journey that takes much longer. The only explanation is that their rapid motion slows down their "internal clock," giving them a long enough life to pass all the way through the atmosphere. Later experiments showed different rates of decay for muons (and also pions) rotating on the outer or inner parts of a spinning disk. Particles near the center move much more rapidly than those farther out and have a longer lifetime.

Another dramatic confirmation of time dilation came in 1972, when physicists reported a test of relativity conducted by flying atomic clocks in jet planes. The flying clocks slowed down, just as Einstein's analysis predicted. (In that case, the time-changing effects of general relativity had to be factored into the analysis as well.)

So Einstein showed, in essence, that the Chicago song was partly right—nobody really knows what time it is. It depends on how you are moving. Neither Einstein nor Chicago, however, posed a similar question about time that a lot of physicists care about today: Does anybody really know *which* time it is? And they aren't talking about time zones. They are talking about the possibility of the second time dimension that Brian Greene mentioned in *Frequency*.

TIME AS A DIMENSION

It is now nearly a century since Einstein predicted time dilation. During that time physics has made enormous progress in understanding motion, energy, matter, and force. Much less depth of understanding has been achieved about the nature of time. I think time still holds some surprises. It's still fair to say that nobody really knows what time is. Consequently there's plenty of speculation about the nature of time, the arrow of time, and the possibility that time could somehow manifest itself in more than one dimension.

A few years after Einstein's special relativity paper, his former math teacher at Zurich, Hermann Minkowski, developed the mathematical treatment of special relativity further, adding time to the three dimensions of space as an equal partner. "Henceforth space by itself, and time by itself, are doomed to fade away into mere shadows, and only a kind of union of the two will preserve an independent reality," Minkowski declared in 1908.[10] The idea of time as a fourth dimension was not exactly new, though. H. G. Wells used it in his science fiction novel *The Time Machine*, and you can find earlier allusions to similar notions if you search seriously enough.[11] But Minkowski, using Einstein's relativity, showed how to make the idea of time as a dimension mathematically precise. It seemed the only way to allow physics to describe events through space and time self-consistently. So physics from then on took place in an amalgam called spacetime, three dimensions of space and one of time. Physicists describe the number of dimensions in a shorthand notation for what they call the "signature" of spacetime: (3,1) (meaning three dimensions of space, one of time).

Over the years, as we've seen, other physicists attempted to fool around with spacetime by adding other dimensions—but almost always just dimensions of space. In the 1980s, though, the idea of additional time dimensions began to creep into the literature, with a mention by the famous Russian physicist Andrei Sakharov and a few others here and there.

Not that anybody noticed. I had never encountered the idea of an extra time dimension in anything remotely newsworthy. But in 1996, when writing about the then brand-new M theory, I heard about "two times" from Michael Duff at Texas A&M. In discussing the signature of spacetime preferred by M theory, Duff mentioned that it was mathematically plausible for more than one time dimension to fit into the equations. In fact, he said, he and a colleague had examined the question in 1988. Applying certain considerations of supersymmetry and other plausible restraints, they had shown which signatures of spacetime remained mathematically consistent possibilities. To their amazement, they found that some scenarios with two time dimensions seemed to make perfect sense—mathematically, at least. For years, this finding was an unremarkable curiosity, but it may turn out to have been a subtle clue to the need to recount the number of dimensions that time has to offer.

TWO TIMES

Adding a dimension of time is a new trick to teach old physicists. Most of them are happy enough to add dimensions of space. But adding a dimension of time is more controversial. Some physicists think it makes no sense. But others think it's the only way to make sense about the latest findings on the frontiers of space and time.

In 1996, Duff could offer no good ideas for explaining what a second time dimension would mean if it did exist. But he noted that the possibility had been taken seriously by some other physicists in connection with M theory. In particular, Cumrun Vafa's F (for father) theory described nature with 10 dimensions of space and 2 of time.

I e-mailed Vafa, inquiring about the meaning of *father* in this context. "My own thinking was if M theory is the 'mother of all theories' as one proponent of it declared, F theory would be the 'father of all theories,' making the relation more politically correct!" Vafa wrote back.[12] "This is not just playing with words," he continued. "In fact the role M-theory and F-theory play in explaining new results in string theory are very much like a cooperative endeavor—(string theory is of course the offspring!)." (If you like, Vafa said, you could have the M stand for male and the F for female.)

As for the second time dimension, its meaning was not exactly clear to Vafa, or to anyone else. A second time dimension might sound like good news for people who are very busy, but the physicists I ask usually get twisted tongues when trying to explain what it would actually mean in real life.

"A hidden time dimension is much more bizarre than a hidden space dimension," said John Schwarz. Edward Witten contended that the second time dimension in F theory merely provides a useful mathematical tool without physical significance. Andy Strominger concurred that the second time dimension seemed to be a mathematical convenience, but he wasn't so sure what to make of it. "You put it in with the right hand and take it away with the left," Strominger told me. "So far it's clearly a calculational trick. But it's a calculational trick that works so well that one suspects there's something more behind it."[13]

Vafa, though, objected to characterizing the second time dimension as merely mathematically useful but without physical meaning. Doubting the physical significance of F theory's second time dimension may be ignoring lessons from history, he suggested.

"Objects which . . . resemble 'abstract mathematical constructions' become more 'physical' when we gain more insight into them," he pointed out.[14] And that's why it makes sense to take the idea of a second time dimension seriously. The whole history of prediscovery shows that mathematical reasoning has often led the way to new physical understanding. In the mid-1960s, many scientists thought that quarks were just convenient mathematical fictions. Now their reality is unquestioned. Even in the early days of superstring theories, the use of extra spatial dimensions seemed to some observers to be merely a way to make the math work out. Now everybody works under the assumption that the extra space dimensions are physically real.

Two time dimensions are simply what you need to make sense out of certain versions of string theory, Vafa contends. Maybe a second time direction seems odd because nobody knows where to look for it—it might come into play only in strange places, perhaps at the center of black holes. So it was too soon, he said, to dismiss the notion that a second time dimension could somehow be real. "As to what that would mean," says Vafa, "I could only say that time will tell."

So far time hasn't told anybody very much. The idea of a second time dimension hasn't grabbed the spotlight among efforts to understand M theory and the relationship of space and time to reality. But the idea hasn't gone away, either. Papers on the second time dimension still turn up from time to time—some advocating the idea, others critiquing it. (It is, after all, one of those ideas that might turn out to be wrong.)

If it's not wrong, though, the key to understanding a second time dimension would be in figuring out why, if it exists, nobody has noticed it. And why it doesn't mess up the world as we know it. For as University of Pennsylvania physicist Max Tegmark has pointed out, it's hard to reconcile a second time dimension with the existence of life.

In a paper he published in the journal *Classical and Quantum Gravity*, Tegmark pointed out that the existence of observers in the universe requires three qualities: complexity, stability, and predictability. That may explain why the universe has only three noticeable space dimensions. A universe with fewer than three space dimensions would not allow enough complexity to produce life. And more than three dimensions would make stable planetary orbits impossible—a planet would either sail off into space forever or smash itself into its star. Atoms would not hold together if their particles were given the freedom of extra dimensions to roam around in. "This means that such a world cannot contain any objects that are stable over time, and thus probably cannot contain stable observers," Tegmark wrote.[15]

A second dimension of time, even though logically possible, could pose equally serious problems, Tegmark believes. True, maybe a second time dimension could exist but go unnoticed. "There is no obvious reason for why an observer could not nonetheless perceive time as being one-dimensional, thereby maintaining the pattern of having 'thoughts' in a one dimensional succession," Tegmark wrote. But there could still be some strange consequences. Any individual would follow a single timeline but might occasionally run into someone moving through spacetime along a different time dimension. If it were a romantic encounter, it could only lead to heartbreak. "If two . . . observers that are moving in different time-directions happen to meet at a point in spacetime, they will inevitably drift apart in separate time-directions again, unable to stay together," Tegmark pointed out.

Such fleeting encounters would be unlikely, though, because life would be rare in a two-time universe. With an extra time direction, the subatomic particles that make up matter would easily disintegrate—in other words, everything would be radioactive. Matter would be stable only in very cold regions, which would limit life to places like Antarctica or Wisconsin. Even worse, extra time dimensions make it impossible to predict the future, Tegmark's analysis shows. With two times, equations describing motion would differ from the usual ones in a way that would make prediction impossible.

"If an observer is to be able to make any use of its self-awareness and information-processing abilities, the laws of physics must be such that it can make at least some predictions," Tegmark wrote. If such predictions are impossible, then "not only would there be no reason for observers to be self-aware, but it would appear highly unlikely that information processing systems (such as computers and brains) could exist at all."[16]

Since there are plenty of computers and brains around today, maybe there are no extra dimensions of time or space. On the other hand, maybe other dimensions exist but in such a way that they don't cause trouble. In string theory, of course, the extra space dimensions are small enough, or isolated enough, to prevent serious problems. Perhaps a similar explanation applies to an extra time dimension—it could be "compactified" into closed curves, so that when you pass into another time dimension, you travel through a very brief loop, too brief to notice. Or perhaps the extra time dimension can be explained within the brane world scenario. Maybe the "bulk" space between branes contains extra space and time dimensions in which our three-brane, with one time dimension, is embedded. We don't notice the extra time because we can't go to the space where it operates.

Itzhak Bars, of the University of Southern California, has written a series of papers expounding on "two-time physics," suggesting that a second time dimension might somehow be "suppressed," sort of like the way the holograms on credit cards display what appears to be a 3-D image on a 2-D surface.

In one interesting paper, Bars and colleague Costas Kounnas of CERN argues that the familiar three-plus-one dimensional universe could have emerged from a stranger universe with up to 11 dimensions of space and as many as 3 dimensions of time. Perhaps, Bars and Kounnas propose, the big bang started only part of the universe expanding—the familiar 3 dimensions of space—while other dimensions remained small and compact. One time dimension might have taken the big-bang route while the others went in another direction.

"A plausible scenario is that one of the timelike dimensions goes along with the expanding universe and the other goes along with the compactified one," Bars and Kounnas wrote.[17]

It's hard to know what to make of ideas as wild as these. For guidance I always go back to Duff, one of those valuable sources who grasps the big picture and offers even-handed assessments. He finds the papers by Bars and most others on the issue "not very compelling." But he expresses intrigue at an approach by the British physicist Christopher Hull. Hull's work focuses on understanding extra times from the viewpoint of duality.

DUAL TIME

Duality, you'll remember from Chapter 8, puts the understanding of the physical world in an entirely new perspective. It changes the very way that the notion of reality is defined. Duality is one of the most profound—and for most people, confusing—ideas of modern physics. And yet at its most basic it captures a message of utter simplicity: what's real depends on how you look at it. Sure, a house is real, but what does it look like? That depends on whether you view it from the front or the back. Sure, an electron is real, but is it a wave or a particle? That depends on what sort of experiment you design to detect it. Sure, your theory of the universe works pretty well. But it's not the only theory that works well, and another one, in some cases might work better, even though other times it works worse. As Niels Bohr used to say, there are two kinds of truth, trivial truths and great truths. The opposite of a trivial truth is obviously false. The opposite of a great truth is another great truth. The dualities of physics are great truths.

When it comes to time, the idea of duality may tell a great truth. Maybe the universe has two times, from one point of view. From our point of view, though, there's only one time. The two viewpoints are dual to each other—flip sides of a coin.

In string theory, it's clear that duality has the power to alter the number of apparent dimensions of space, as we saw in Chapter 8. A supermembrane can wrap itself around a space dimension, kind of like Handi-Wrap around a hot dog, except there's no hot dog, just a dimension of space. Suppose that dimension shrinks. The Handi-Wrap tightens around it, and sooner or later the Handi-Wrap looks more like a string than a membrane, and the dimension it surrounded seems to have disappeared. A theory with branes in 11 dimensions now looks like a theory of strings in 10 dimensions.

Chris Hull's insight into the time problem is that dualities can do the same thing with time that they do with space. In switching between dual descriptions, such as when moving from the realm of weak coupling to strong coupling, maybe more than the number of space dimensions can change.

"Remarkably, it turns out that dualities can change the number of time dimensions as well," Hull wrote in one paper, "giving rise to exotic spacetime signatures. The resulting picture is that there should be some underlying fundamental theory and that different spacetime signatures as well as different dimensions can arise in various limits."[18]

So whatever the "fundamental" theory of the universe is, it should not specify one preferred spacetime signature. "Any attempt to formulate M theory or string theory as a theory in a given spacetime dimension or signature will be misleading," Hull contends. "In particular, the theory underpinning all these theories . . . cannot at a fundamental level be a theory in 10+1 dimensions, as it has some limits which live in 9+1 dimensions and others that live in 9+2 or 6+5 dimensions."[19]

In other words, the (3,1) signature of ordinary spacetime is just the one that seems most convenient from the human viewpoint. Nature may encompass other signatures, some with more than one time dimension, that are just all dual versions spawned from some grander concept. There is no such thing as just "spacetime" but instead a whole class of different spacetimes, ultimately equivalent, just as the five versions of string theory turned out to be equivalent in the end.

"We have seen that different spacetimes related by dualities can define the same physics, so that the notion of spacetime geometry cannot be fundamental," Hull asserts. Or as my friend K. C. likes to say, at the most fundamental of levels, "Space and time are toast." Spacetime should be a derived concept, built from something else. At the moment, though, nobody knows what the something else is.

In doing away with the idea of spacetime as fundamental, Hull sees a parallel with Einstein's relativity theories. The different frames of reference of special relativity (expanded by general relativity to include any set of spacetime coordinates) are all equally valid for describing nature. The frame of reference you use depends on the frame of reference you inhabit. In the same way, many different spacetime signatures may turn out to be equivalent, and we organize physics based on the signature that seems most sensible from our point of view.

"Two dual theories can be formulated in spacetimes of different geometry, topology and even signature and dimension," Hull notes. "And so all these concepts must be relative rather than absolute, depending on the values of certain parameters or couplings, and such a relativity principle should be a feature of the fundamental theory that underlies all this."[20] Duality may describe the second coming of Einstein's relativity principle, in a new, more powerful form, with a vengeance.

Many of the implications of Einstein's relativity seemed very strange, as they applied to realms of phenomena far from ordinary experience. In a similar way, the duality idea could explain the strangeness of a second time dimension. Spread all the possible spacetimes out in a room, and you'd find that the one we inhabit is tucked off in a corner somewhere. In that corner of all possible spacetimes, an extra time dimension poses no problem. If the "real" theory has 9 space and 2 time dimensions (signature 9,2), it can be rewritten in our corner as a dual theory with 1 time and 10 space dimensions. In other corners of all possible spacetimes, where we do not live, the (9,2) signature might be much more natural.

"Some corners are stranger than others, but in any case we can only live in one corner . . . and there is no reason why other corners might not have quite unfamiliar properties," says Hull.[21]

In other words, our grasp on reality is limited. Appearances can be deceiving. Science is about finding out what lies behind the appearances. And what lies behind might just be a second dimension of time. After all, if a second dimension of time makes the math work, there just might be something to it. As we've seen, math can predict some very strange things that later turn out to be discovered out there in the real world. It's time to try to understand how math is able to do it.

EPILOGUE

I am a little piece of nature.

—Albert Einstein

When the young journalist Walter Lippmann wrote a book in 1922 called *Public Opinion*, he titled the first chapter "The World Outside and the Pictures in Our Heads." There was, he observed, a big difference between the two.

"What each man does is based not on direct and certain knowledge, but on pictures made by himself or given to him,"[1] Lippmann noted. "The world that we have to deal with politically is out of reach, out of sight, out of mind. It has to be explored, reported, and imagined."[2]

Such truths about the political world, Lippmann would no doubt have agreed, apply to the natural world as well. For the pictures of the natural world that science provides are not snapshots of a naked reality, but artistic renditions that attempt to capture reality's essence. Reality always appears distorted by the imperfection of human senses

255

and the filters of the human mind. Scientists interpret reality's shadows, helping to guide life's journey through nature's jungles.

But if nature is separate from science's pictures of it, on what foundation does science build its claim to reflect some "truth" about reality? As its critics often point out, science's version of the "truth" is always changing. From the ancient Greeks, to Newton, to Einstein, to today, science's comprehension of the universe has evolved, mutated, matured, and been reborn. New nuances constantly emerge, and radical departures from textbook dogma occasionally appear.

Yet while scientists agree that their knowledge is tentative, almost all nevertheless insist that it captures some truth about an objective reality. Many of science's critics, on the other hand, dispute the very idea of a truth to be captured. They argue that "reality" is beyond the reach of the human mind. The laws of physics are not really deep truths about nature, the critics claim, but mere agreements among a community of scientists about how to talk about nature.

In responding to such claims, many scientists are content to side with the seventeenth-century German philosopher Gottfried Wilhelm von Leibniz. If we can apply reason to the world and not be deceived, Leibniz said, the world is "real enough." And from a practical viewpoint, science's success in applying reason to reality is spectacular.

But I think a more compelling case can be made for science's ability to grasp an independent reality. That case hinges on the success of mathematics in describing the physical world and on math's ability to enable prediscoveries.

Somehow, humans have been able to discover the laws governing nature, in the form of symbols and the rules for combining them. Those symbols, imaginatively manipulated, have foretold the existence of strange objects and phenomena—antimatter and quarks, neutrinos and black holes, radio waves and vacuum energy, the expansion of the universe and the curvature of space. Most scientists believe that math's success in this regard signifies something deep and true about the universe, disclosing an inherent mathematical structure that rules the cosmos, or at least makes it comprehensible.

Nevertheless, scientists have a tough time explaining how it is that math works so well. As Eugene Wigner expressed it so eloquently four decades ago, "The enormous usefulness of mathematics in the natural sciences is something bordering on the mysterious and there is no rational explanation for it."[3] And Wigner's essay did not even address the power of mathematics to discover new things in advance of any physical clues to their existence. It's those prediscoveries, I believe, that provide science's best clue to the existence of a reality that science can perceive. Is there something real out there? Math's ability to divine the presence of strange matters in the universe argues strongly that yes, there is.

Yet a recent popular book seems to contradict that idea completely. As I mentioned in the Introduction, two cognitive scientists argue that math is merely a human invention that has nothing inherent to do with any external reality.

MATHS R US

George Lakoff, of the University of California, Berkeley, and Rafael Núñez, of Berkeley and the University of Freiburg, in Germany, see a world ruled not by math, but by the human brain. Whatever might be "real," they write, human knowledge depends solely on the brain and its own ways of finding things out. Math is not a discovery about the external world, but an invention rooted in metaphors linked to human thoughts, sensations, and actions.

"Where does mathematics come from?" Lakoff and Núñez ask. "It comes from us!" they answer in their book, *Where Mathematics Comes From*. "We create it, but it is not arbitrary," they write. "It uses the basic conceptual mechanisms of the embodied human mind as it has evolved in the real world. Mathematics is a product of the neural capacities of our brains, the nature of our bodies, our evolution, our environment, and our long social and cultural history."[4]

All mathematical ideas, Lakoff and Núñez contend, are elaborate metaphors. Those metaphors are drawn from real-world experience and then linked and mixed to guide mathematical practice. Basic principles of arithmetic, algebra, trigonometry, mathematical logic, and other math fields all rely on metaphorical reasoning.

Such metaphors underlie rigorous systems of deduction and calculation. Arithmetic, for example, can be envisioned as movement along a path with markers placed at equal intervals, the metaphorical basis for the concept of a number line. Other metaphors can be identified to illustrate the ideas underlying more advanced math. Therefore, Lakoff and Núñez conclude, math is a mere human invention, a systematic way of capturing the way the brain sees the world.

In their book, Lakoff and Núñez declare that math succeeds in science only because scientists force it to. The fit between mathematics and the regularities in the world is all in the mind of the mathematician-scientist, not in the physical universe outside.

At a meeting in San Francisco, in February 2001, I heard Lakoff defend this view. He hedged a little on the issue of whether math really does exist in the world apart from in the human brain. "The only mathematics that we know is the mathematics that our brain allows us to know," he said.[5] So any question of math's being inherent in physical reality is moot, since there is no way to know whether it is or not. "Mathematics may or may not be out there in the world, but there's no way that we scientifically could possibly tell," he argued.

Well, I suppose he might be right. But I doubt it. That view surely does not correspond to what many great scientists think about the issue. I appreciate much more the view expressed by Heinrich Hertz, who discovered electromagnetic waves after James Clerk Maxwell prediscovered them, in commenting on Maxwell's theory.

"It is impossible to study this wonderful theory without feeling as if the mathematical equations had an independent life and an intelligence of their own, as if they were wiser than ourselves, indeed wiser than their discoverer, as if they gave forth more than he had put into them," Hertz said.[6]

Murray Gell-Mann, prediscoverer of quarks, makes a similar point. He views efforts to find the ultimate theory of nature's particles and forces not as a construction job, but as an exploration. "It seems that this whole theory is lurking there in some mathematical space," Gell-Mann said during a talk at Caltech in 2000. "It is there to be found. . . . The search for it appears to be a process of discovery, not invention. You are not adding bells and whistles in an effort to fit some empirical facts. You are gradually finding out what that preexisting self-consistent structure is."[7]

I don't think Gell-Mann would like Lakoff and Núñez's book.

Still, even if math resides in the physical world, that fact doesn't solve the big mystery. Sure, humans can observe the universe and then find equations that capture patterns in what goes on out there. You can watch planets move through the sky, for example, and detect regular features that allow you to infer equations that describe the path of any of them. That's what physics is all about—finding the formula that applies in general to everything within a class of phenomena. Humans, in other words, can translate reality into squiggles on paper.

But how is it that those squiggles then reveal aspects of reality that had never been observed? It's like translating Euripides from Greek into English and getting the works of Sophocles as a bonus. How do you get more out of the equations than what you put in? How is prediscovery possible?

INTIMATE WITH NATURE

During the months I spent writing this book, I posed that last question to many physicists. The quickest answer came from Rocky Kolb.

"A lot more things are proposed than ever turn out to be true," he said. "If enough people are proposing enough things, some of them are bound to be right. A lot of it is luck."

No doubt that is part of the answer. If that were the whole story, though, I could end the book now. But Rocky acknowledged that there is more to it than that. In the game of prediscovery, some people are suspiciously lucky.

"When they do it more than once, it's like winning the lottery twice in a row," Rocky says. "You start to suspect there's something funny about the balls in there. Obviously they have some insight or imagination that I don't see, that I don't have."[8]

Somehow the great prediscoverers have an intimate relationship with nature, or have an enhanced intuition about how the universe operates. Hertz said something like that about Maxwell and his equations. "Such comprehensive and accurate equations only reveal themselves to those who with keen insight pick out every indication of the truth which is faintly visible in nature," Hertz said.[9]

Maxwell developed his equations through contemplating actual physical devices. His appreciation of the way the physical world worked guided the search for the right math. In a similar way, Alexander Friedmann's intimacy with the atmosphere guided his interpretation of Einstein's equations applied to the universe. Friedmann knew the weather; he knew how to translate the squiggles on paper into wind speeds and air pressures. His experience relating equations to the atmosphere surely prepared him to better comprehend the relationship of Einstein's equations to all of space.

Einstein himself, the greatest prediscoverer of them all, snared secrets from nature as though he somehow had illegal access to inside information. Gerald Holton summed it up this way, using Einstein as the archetype for scientific genius: "There is a mutual mapping of the mind and lifestyle of this scientist, and of the laws of nature."[10]

In other words, Einstein's mind somehow embodied a map of the natural world. But when Einstein said he was "a little piece of nature," he was speaking not only for himself. Everybody possesses a mental model of reality. Your brain consults that model when directing your actions, using it as a guide for navigating in the world. Why does this work? Because people aren't mere observers of a real world outside. People and their brains are part of the world that science tries to discover and explain.

From this point of view, Lakoff and Núñez may not be quite as far off base as they seem. In fact, they may have latched on to part of the solution, even while denying the existence of the mystery. They insist, you'll recall, that math is merely a human invention and that any usefulness for describing nature arises simply from the fact that this is the way our brains work. But if math is merely a human invention, then how can it tell us about things in the real world that have not yet been seen? Perhaps because our brains are part of the physical world, too. Maybe the math that brains invent takes the form it does because math had a hand in forming the brains in the first place (through the operation of natural laws in constraining the evolution of life).

Einstein was able to exploit this intimate relationship with nature more successfully than most humans. But how did he do it—let's say, how did he get in touch with his inner self? It looks to me that he accomplished what he did by recognizing the power of ideas, especially of ideas in which he could find the principles that nature puts on a pedestal. Einstein showed that by understanding some one important thing about reality, and then by elevating that insight to the status of a principle, you could discern many previously unknown things about reality.

At this point, I'm struck with how opposite this picture is to the one I drew in my previous book, *The Bit and the Pendulum*. In that book I identified great technologies as the inspirations for progress in science. The mechanical clock provided the metaphor exploited by Newton to generate the mechanical "clockwork" view of a universe governed by force. The steam engine inspired Sadi Carnot to lay the foundations for thermodynamics, a science picturing a universe ruled by energy. In the twentieth century the computer infiltrated scientific culture, driving new investigations of natural processes that can be described in terms of information storage and processing.

Now it looks as though I'm saying the opposite. Progress in science—and especially in prediscovering unknown things— is driven not by grand technologies, but by grand ideas. In fact, I see three main ideas that have taken turns inspiring the last two centuries of progress in physics and cosmology. In the nineteenth century (and before), the guiding idea was "geometry." In the twentieth century, the most fruitful principle was "symmetry." And the idea of the twenty-first century, I believe, will be "duality." Examining those ideas goes a long way to helping understand the mystery of prediscovery.

GEOMETRY

The idea behind geometry is old and simple: insight into nature can be gained by building an edifice of logical deductions based on "self-evident" axioms. The tricky thing about geometry, of course, is the apparently innocent idea of *self-evident*. It seems to me that *self-evident* is just a way of saying that that's the way things look like in the world. Geometry, in other words, did not originate from pure thought but from human experiences with nature. Long before Euclid got around to codifying its logical foundations, geometry developed from solving the practical problems encountered by Egyptian surveyors.

Euclid compiled a lot of existing geometrical knowledge and showed how its insights could be derived from a small number of definitions and propositions. To the extent that those propositions merely reflected common experience, geometry was not really an independent, strictly logical system divorced from any certain connection to reality. It was rooted in reality, or at least in reality as it appeared to the early geometers. In particular, Euclid's famous fifth postulate could only have reflected physical considerations. Parallel lines never meet, Euclid said. Or the shortest line connecting two parallel lines meets them at right angles. It's all the same thing. But there's no way to prove it—there's no strictly logical reason why it has to be true, even given the other axioms. The only way you would know it's "true" is by drawing lines and measuring angles in the real world.

Nevertheless, nineteenth-century thinkers thought Euclid's geometry revealed truths about reality. Here's the odd thing, though. Euclidean geometry—the geometry that incorporates the lessons from real life—turns out to be the wrong geometry for describing real life. The non-Euclidean geometries of the nineteenth century did away with Euclid's fifth postulate, sticking to a more rigorously logical-mathematical approach, uncontaminated by physical appearances. Gauss, Lobachevsky, Bolyai, and especially Riemann recognized the limitations of experience in grasping reality. It was the divorce of geometry from experience that led to the great prediscovery of space (or spacetime) curvature. Pure mathematical reasoning gave back something that wasn't put into it.

SYMMETRY

Geometry, of course, is intimately connected with the notion of symmetry. Moving and rotating and reflecting geometrical figures leave the deductions about them unchanged—a good thing, or otherwise Euclid would have been wasting his time. Einstein exploited the symmetry in geometry to explain how the laws of nature remained unchanged from different points of view.

In his general theory of relativity, Einstein reaped the benefits of Riemann's insights, showing how the new geometry could describe gravity in such a way that everybody observes the same law of gravitation. That accomplishment was rooted in Einstein's success with symmetry in special relativity. Maxwell's equations, in Einstein's eyes, captured something essential about reality. It would be heresy to alter those equations merely to accommodate the motion of some observer at a constant rate of speed along a straight line. There must be a symmetry that keeps the laws the same even when motion changes. Einstein built special relativity on that notion.

The success of special relativity, and then of general relativity a decade later, revealed symmetry's power to twentieth-century science. Later physicists, building on Einstein's inspiration and the work of others like Emmy Noether, Hermann Weyl, and Eugene Wigner, constructed a whole outline of existence from the blueprints contained in mathematical symmetries. The symmetry approach brought science to an incredibly deep understanding of nature's laws, in the form of the Standard Model of particles and forces, well before the end of the twentieth century.

Just as the symmetries recorded their greatest victories, though, progress stalled. Then, in the twentieth century's last years, a new idea for the new millennium arose, with the promise of sustaining the quest for the ultimate understanding of the universe. That idea encompasses a special kind of symmetry that goes by the name of duality.

DUALITY

Duality at first glance seems to be an utterly simple idea. The back of a house looks different from the front, but it's the same house. So what? Such a duality seems to offer a rather trivial insight. As superstring theorist Brian Greene points out, many dualities *are* trivial—say, Einstein's theory presented in Chinese or in English.

True, they are different-looking ("dual") views of the same theory. But if you learned Einstein's theory in English, you will discover nothing more about it by reading a Chinese translation.

In other situations, duality runs deeper—like the duality between ice and steam. Suppose you are a scientist in a primitive culture, trying to figure out what steam is made of. It won't be easy—it's hard to keep steam contained, and you're likely to scald yourself. But then you discover that ice and steam are just two very different appearances of the same substance. Ice is easier to study. You can experiment with it, measure things about it, and ultimately figure out that it's made of molecules containing hydrogen and oxygen. You can learn things you never knew about steam by studying ice, thanks to the ice-steam duality.

In the same way, theories of the universe can look very different yet be identical on some deeper level. And one of those theories might be much easier to use. This is precisely the situation in certain string theories. When strings interact strongly, the math describing them is very hard to do. But when strings interact weakly, the math is much easier. To solve hard problems with strongly interacting strings, you can use the dual theory, with weakly interacting strings.

Recently physicists have explored another profound duality, discovered by Juan Maldacena at Harvard. (He is now at the Institute for Advanced Study in Princeton). He found that the physical description of a volume of space (let's say, for example, the interior of a black hole) can be equally well represented by the boundary of that space (the surface, or "horizon" of the black hole). The two descriptions are dual. It's like saying you could describe all the three-dimensional objects in a room merely by looking at a wall—a two-dimensional surface containing three-dimensional information. (Think about it: the universe might be like a room with walls covered by mirrors.)

Physicists are still trying to figure out what all of this means. But at least it's clear that the search for an ultimate physical theory will not produce one.and only one picture. Different points of view reveal different pictures.

"The main lesson of recent progress in string dualities has been the recognition of the existence of different viewpoints on a physical theory all of which are good for answering some question," says Harvard's Cumrun Vafa. "It is this democracy of physical descriptions (not the superiority of one over the other) which is the lesson of string duality."[11]

Duality therefore expresses a special kind of symmetry, a symmetry between theories. But duality symmetries still share much with the symmetries of geometry and the symmetry of the twentieth century embodied in gauge theories. Geometry, symmetry, and duality are themselves just different aspects of the same thing. They all point to an underlying sameness, an underlying identity that scientists struggle to discern. Geometry describes symmetries of space. In Einstein's hands, symmetry transformed space into spacetime, an arena where gauge symmetry could describe particles and forces. Duality symmetries show now that spacetime is not the final answer. But nobody yet knows what to replace it with.

At least the idea of duality has helped me immensely in reconciling my previous book with this one. Obviously, the two books are dual to each other.

UNITY AND HARMONY

Geometry, symmetry, and duality have taken science a long way. Taken together, they also explain a lot about why an ultimate picture of reality has been so elusive. It's because things are not always what they seem. Euclidean geometry looks right, but only if there's no matter around to warp space. Newton's laws of motion seem fine, unless you happen to be moving really fast. J. J. Thomson won a Nobel Prize for proving the electron is a particle; then his son George won a "dual" Nobel 31 years later for showing that the electron is a wave.

At the same time, though, geometry, symmetry, and duality show how to find an underlying sameness that makes the world consistent with itself. Riemannian geometry makes general relativity work, so that you can describe the universe with whatever system of coordinates you choose. The symmetries of special relativity translate the description of slow-moving or fast-moving bodies into the same language, keeping nature's laws the same for everybody. Duality permits string theory to appear in many disguises, but also allowed Edward Witten to realize that the five superstring theories were just five doors to the same M-theory house. All this progress points to an underlying unity in a hard-to-see reality. And that unity is our strongest hint of objective reality, however dimly we perceive it.

This picture is clearly in harmony with Einstein's longtime quest to find the unified theory of gravity and electromagnetism, a quest that has mutated into today's search for a "theory of everything." And it's a picture in harmony with the writings of the seventeenth-century Dutch philosopher Baruch Spinoza, who deeply influenced Einstein's attitudes on these issues. Spinoza's God, Einstein wrote, "revealed himself in the harmony of all being." Understanding that harmony was Einstein's lifelong goal.

In Spinoza's view, all the variety in the perceived world ultimately stemmed from an essential underlying harmony.[12] But science can't grasp that reality as a whole. Science must deal with the fragments of nature accessible to human perception. To Spinoza (and, apparently, to Einstein), all the fragments are pieces of an infinite puzzle. The whole puzzle is a unified "substance" at reality's foundation. Spinoza's substance is only imperfectly perceptible, as humans have access only to its manifestations, not its inherent unity.

I think this point of view explains prediscovery. At least I think such a case can be made—and has, in fact, been made pretty well by the French mathematician Henri Poincaré. During the first years of the twentieth century, Poincaré was one of France's great scientific popularizers. He thought deeply and wrote clearly about the relationship of science and mathematics to reality. And while he never (as far as I've been able to find) addressed the idea of prediscovery directly, things he said in different places offer a clue to what he would have said about it had he been asked. My reconstruction of what he would have said goes something like this:

1. You can't add apples and oranges.
2. If it gets you where you want to go, the map must be right.
3. If you used math to make the map, the world is not a fruit salad.

Of course, Poincaré said these things a little more eloquently.

"It might be asked, why in physical science generalization so readily takes the mathematical form," he wrote in *Science and Hypothesis*. "It is because the observable phenomenon is due to the superposition of a large number of elementary phenomena which are *all similar to each other*."[13] (Spinoza might say the elementary phenomena are all expressions of his harmonious substance at the root of reality.) In other words, math works because it deals with similar things.

"Mathematics teaches us, in fact, to combine like with like," Poincaré wrote. "Its object is to divine the result of a combination without having to reconstruct that combination element by element."[14] You don't have to count to 100 five times, for instance, to learn that 5 times 100 equals 500. But for math's shortcut to work, you have to combine like with like. Multiplying 5 apples times 100 oranges would be senseless. And so math's success in science implies that science deals with "likes."

The second point, about maps getting you where you want to go, refers to the use of equations to represent reality. We can't see the "true" underlying reality directly, so we substitute images for the "true" objects. We assign symbols that can be manipulated mathematically to represent these images. The success of the math signifies that the relations between the images conform to the relations between the underlying "real" objects.

In other words, equations represent relationships. Using equations you can construct a mathematical map that helps you find your way through nature.

For example, Poincaré suggested, consider physical concepts such as motion or electric current. You can assign symbols to stand for these concepts. Then you can deduce physical consequences by manipulating the symbols according to the formal rules of math. By definition, those formal rules have nothing to do with reality. But those rules do succeed in making accurate physical deductions. The rules of math must therefore contain something true about the relations between the physical things in the real world. "If the equations remain true, it is because the relations preserve their reality," Poincaré wrote.[15]

To sum it up, Poincaré says math works because it deals with like things. Since math works for science, science must be dealing with like things, not a fruit salad of disconnected realities. Math's success suggests a deep underlying "likeness" in the universe, a simplicity, or unity, which reveals a connectedness, or universal set of relationships, connecting absolutely everything. It is math's ability to express those relationships that allows science to identify truths about reality before they are observed. Since everything is harmoniously connected, observing part of the whole can tell you about other parts you haven't seen yet.

Some theories are, of course, better than others at facilitating prediscovery. Observations are always approximate, and theories based solely on observations may therefore not capture precisely the right relationships. Those theories might explain observations over a restricted range of conditions but fail under other circumstances.

This is true even of some very good theories, such as Newton's explanation of gravity. It fails when gravity is very strong.

Other theories, though, capture more of the whole of the universe and preserve more nearly all of the underlying connections and relations. These theories, like Einstein's, provide a vastly greater range of insights into the universe, and lead to prediscoveries.

Poincaré understood how this process works. Imagine, he says, a chart depicting all "the variations of the world."[16] At each point in time everything in the universe is in a particular arrangement. At the next instant the arrangement will be slightly different. (The differences from one instant to the next would be the result of the combined operations of all the laws of physics.) A graph of those changes over time would take the shape of a curve. A good mathematician could figure out an equation to describe that curve. With that equation, then, we could extend the curve to predict the future of the universe or ascertain the past.

But earthbound mathematicians can never see the whole curve. Human theories are always based only on one arc, one piece of the universal curve. Two theories based on different arcs might deduce different equations to describe the whole curve. (Quantum mechanics and general relativity describe different aspects of reality exquisitely well, for example, while appearing to be incompatible.)

However, Poincaré notes, a greater intellect, or a similar intellect with a wider field of view, could perceive the region between these two arcs and construct a better equation. That equation could describe not only both arcs but also the part of the curve in between. And sometimes human scientists can figure out that better equation before they see the whole curve. If they get the right equation, it will then tell them things about regions of the curve that have not yet been measured. In that way the math can reveal things about the physical world that haven't yet been seen.

Historically, this is almost exactly what happened a century ago when Max Planck introduced quantum theory to the world. In 1900 there were two formulas for predicting how bright different colors of light would be when emitted from a hot oven (or "black body" cavity). One formula worked well if wavelengths of the light were short, toward the blue end of the spectrum; the other worked well when wavelengths were long, toward the red end of the spectrum. Planck found the equation that worked for all wavelengths.

And then he figured out what his equation meant. His equation could be true, he found, only if energy could not be divided into smaller and smaller amounts. Energy had to come in packets, or quanta, the way money comes in units no smaller than pennies. This realization launched the quantum revolution. Today scientists know that the world at its foundations is very, very strange, obeying laws that strike many people as bizarre. But it's the way the world is.

Planck's formula explained the spectrum of black body radiation. But that formula told him much more than just the intensity of light at different colors. It told him, once he thought about it, something deep and true about the nature of reality at the most fundamental of levels, something that applied to much more than light coming out of an oven. Somehow the math told Planck about something real. Planck's math revealed something that nature had been hiding. He got more out of his equation than what he put in.

In this regard I think we should remember something else that Planck once noted. "Great caution must be exercised," he said, "in using the word, *real*."[17] Poincaré expressed similar caution. "What this world consists of, we cannot say or conjecture; we can only conjecture what it seems, or might seem to be to minds not too different from ours," he wrote.[18] What we learn about reality are relationships, relationships expressed in mathematics. The objects of reality we think we have discerned are merely images that allow us to visualize the relationships that math reveals.

"The true relations between these real objects," Poincaré commented, "are the only reality we can attain."[19] Equations may show a relationship between motion and electric current, for example. "But these are merely names of the images we substituted for the real objects which Nature will hide for ever from our eyes," Poincaré wrote.

Nevertheless, math's success tells us that the universe is real. Science evolves, though, because our imperfect perceptions of the underlying reality can never capture the entire true picture. One approximate picture after another emerges as new ideas inspire new images. To describe what that reality is like, we have no recourse but to say what it seems to be like, and we naturally choose for comparison those tangible mechanisms and processes that capture and shape our cultural imaginations.

To ancient thinkers, with minds different from ours, the world seemed to be a different place—a series of concentric spheres. In the late Middle Ages, people began to view the universe as a clockwork, inspired in this belief by the cultural importance of the mechanical clock. By the end of the nineteenth century the universe seemed more like a big steam engine (one that was running out of steam), the prime mover of the Industrial Revolution. Nowadays, to many scientists, the world seems a lot like a computer—a point of view clearly inspired by the computer's role as society's dominant machine.[20]

These metaphors are like the images that Poincaré spoke of, capturing essential relations stemming from that unseen reality beyond our senses. This may strike some as unsatisfying, for it seems that what we usually think of as real isn't really real at all. But in the end the situation is just as Lippmann perceived it in politics. The world outside is different from the pictures in our heads, and the "real" world is not exactly the same thing as the pictures in the scientists' heads. As Lippmann wrote in *Public Opinion*, "the real environment is altogether too big, too complex, and too fleeting for direct acquaintance."[21] And scientists are only human.

"Man is no Aristotelian god contemplating all existence at one glance," Lippmann wrote. "He is the creature of an evolution who can just about span a sufficient portion of reality to manage his survival."[22]

NOTES

INTRODUCTION

1. Women seem to be described very harshly by Abbott, starting with their lower status as less than complete polygons. This was Abbott's way of satirizing the ridiculous discrimination against women common in his day.

2. By combining a fourth dimension, time, with the familiar three dimensions of space, Einstein's equations could describe laws of physics that did not depend on how an observer was moving. That was a good thing, because it would be a pretty messy universe if the math to describe it changed whenever you moved. With relativity theory, any observer can establish a frame of reference to describe what goes on, giving an event a set of four coordinates, specifying its time and location in the three dimensions of space. Observers moving in different directions and at different speeds will generally describe things with a different set of coordinates. Einstein's equations make it possible for the laws of nature to stay the same for these different observers, or when the frame of reference of any particular observer changes. Einstein was not explicit about this in his original papers; the first clear enunciation of time as the fourth dimension in relativity theory came from the mathematician Hermann Minkowski.

3. Rosie is a medical writer for the *Los Angeles Times*, where she writes the funniest yet most intelligent health column in the nation.

4. Wigner, E. 1960. The unreasonable effectiveness of mathematics in the natural sciences. *Communications in Pure and Applied Mathematics*. P. 534 in *The World Treasury of Physics, Astronomy, and Mathematics*, T. Ferris, ed. Boston: Little, Brown, 1991.

5. Wigner, p. 536.

6.   Lakoff, G., and R. Núñez. 2000. *Where Mathematics Comes From*. New York: Basic Books, p. 3.

7.   Lakoff and Núñez, p. 344.

STRANGE MATTER

1.   Jaffe, R. L., W. Busza, F. Wilczek, and J. Sandweiss. 2000. Review of speculative "disaster scenarios" at RHIC. *Reviews of Modern Physics* 72:1126. Also available at xxx.lanl.gov/abs/hep-ph/9910333.

2.   Leon Lederman, interview with the author at Fermilab, June 16, 1997.

3.   Gell-Mann proposed the strangeness idea in 1953; later that year the same idea was developed independently by Tadao Nakano and Kazuhiko Nishijima in Japan. A good, brief but more technically detailed account of the origins of strangeness is given in Pais, A. 1986. *Inward Bound*. New York: Oxford University Press, pp. 519-521.

4.   Yuval Ne'eman, an Israeli physicist, independently proposed the same basic idea at about the same time.

5.   Group theory was not obscure to mathematicians, of course. More details on group theory will appear in Chapter 3.

6.   Murray Gell-Mann, lecture in Santa Fe, N.M., September 23, 1999.

7.   Pais, A. 1952. Some remarks on the *V*-particles. *Physical Review* 86:672.

8.   Gell-Mann, interview by the author in Santa Fe, N.M., September 16, 1997.

9.   An electron volt is a unit of energy equal to the amount of energy it takes to boost an electron through a potential of 1 volt. But it is used as a convenient unit of mass in particle physics, reflecting the interchangeability of mass and energy. The mass of a proton is a little less than 1 billion electron volts, or 1 GeV.

10.   Barnes, V. E., et al. 1964. Observation of a hyperon with strangeness number three. *Physical Review Letters* 12(February 24):206.

11.   Gell-Mann, interview by the author in Santa Fe, N.M., September 16, 1997.

12.   Willy Fischler, lecture at Southern Methodist University in Dallas, TX, February 8, 1999.

13.   Brian Greene, conversation at dinner with the author in Ann Arbor, Michigan, July 11, 2000.

14.   Edward Witten, interview by the author in Ann Arbor, Mich., July 10, 2000.

15.   Witten points out that it's possible, perhaps, that current ideas about the big bang will turn out to be wrong. Strictly speaking, the astronomical evidence indicates that the universe was very hot and dense in its youth. It's conceivable that at the very beginning it was cold—conditions under which it might have been possible to create strange quark matter. "This would be a good idea if the big bang were really cold and the heating occurred later, after the quark matter was formed," Witten said. In fact, some scientists have speculated on the possibility of a cold big bang, notably Harvard astrophysicist David Layzer. But the overwhelming consensus of cosmologists remains otherwise. "Quark matter in the early universe is very hard to make, if it's true that the early

universe was hot," Witten says. "What we really know is that today the ratio of photons to baryons (heavy particles such as protons and neutrons) is very high. If that was true in the early universe, then the early universe was a very bad place to make quark matter."

16. Witten, interview in Ann Arbor, Mich., July 10, 2000.

17. At the moment (February 2002) Teplitz is on leave from Southern Methodist University to serve as a senior policy analyst with the White House Office of Science and Technology Policy.

18. Strange quark nuggets might be stable over a wide range of sizes, from baseball-sized chunks weighing a trillion tons down to the size of an ordinary atomic nucleus. Lightweight nuggets might burn up in the atmosphere, displaying themselves as meteors. See Crawford, H., and C. Greiner. 1994. The search for strange matter. *Scientific American* 270(January):72-77.

19. In these calculations, the authors suggest that an initially lightweight strangelet, smaller than a uranium nucleus, actually grows in mass by absorbing neutrons (and maybe some protons as well) from atoms in the air on its way down. See Banerjee, S., et al. 2000. Can cosmic strangelets reach the earth? *Physical Review Letters* 85(August 14):1384-1387.

20. Broderick, J., et al. 1997. Millimeter-wave signature of strange matter stars. xxx.lanl.gov/abs/astro-ph/9706094, June 10.

21. Jaffe, et al., p. 1136.

MIRROR MATTER

1. Pais, A. 1986. *Inward Bound.* New York: Oxford University Press, p. 286.

2. Gamow, G. 1961. *Biography of Physics.* New York: Harper & Row, p. 262. While this story seems consistent with Dirac's character, it's worth keeping in mind that Gamow's anecdotes are not always easy to verify.

3. Heilbron, J., and T. Kuhn. 1969. The genesis of the Bohr atom. *Historical Studies in the Physical Sciences* 1:257.

4. Hermann, A. 1971. *The Genesis of Quantum Theory*, translated by C. Nash. Cambridge, Mass.: MIT Press, p. 157.

5. Dirac, P. 1978. *Directions in Physics.* New York: John Wiley & Sons, p. 4.

6. When Heisenberg showed his math to Max Born, Born informed him that it was merely a reinvention of matrix algebra, a development dating to the 1850s.

7. Dirac, P. 1983. The origin of quantum field theory. P. 44 in *The Birth of Particle Physics*, L. Brown and L. Hoddeson, eds. Cambridge: Cambridge University Press.

8. Dirac, *Directions in Physics*, p. 14.

9. Dirac, *Directions in Physics*, p. 15.

10. For a good explanation in more depth, see Dirac's own discussion in *Directions in Physics*, pp. 11 ff.

11. Dirac, P. 1930. A theory of electrons and protons. *Proceedings of the Royal Society* (London), Series A, 128. P. 1195 in *The World of the Atom*, H. Boorse and L. Motz, eds. New York: Basic Books, 1966.

12.   Dirac, *Directions in Physics*, p. 17.

13.   Oppenheimer, J. R. 1930. On the theory of electrons and protons. *Physical Review* 35. P. 1205 in Boorse and Motz.

14.   Pais, A. 2000. *The Genius of Science*. New York: Oxford University Press, p. 59.

15.   Gordon Kane, conversation with the author in Ann Arbor, Mich., July 11, 2000.

16.   Dirac, P. 1971. *The Development of Quantum Theory*. New York: Gordon and Breach, p. 56.

17.   Dirac, *Directions in Physics*, p. 17.

18.   Anderson, C. 1999. *The Discovery of Anti-matter*. Singapore: World Scientific, p. 25.

19.   Anderson, C. with H. Anderson. 1983. Unraveling the particle content of cosmic rays. P. 140 in Brown and Hoddeson.

20.   This is not true of all particles without charge. The neutron, for instance, has no net electrical charge. But its component quarks do. An antineutron comprises the antimatter counterparts of its three quarks. Instead of two downs and an up, as in the neutron, the antineutron is made of two anti-downs and one anti-up. The charges still add up to zero.

21.   Weyl, H. 1949. *Philosophy of Mathematics and Natural Science*. Princeton: Princeton University Press, p. 208.

22.   Crease, R. P., and C. C. Mann. 1986. *The Second Creation*. New York: Macmillan, p. 209.

23.   Leon Lederman, interview with the author at Fermilab, 1997.

24.   Yang, C. N., and T. D. Lee. 1956. *Physical Review* 104 (October):258.

25.   There is a possibility that some very slight electromagnetic interaction might be possible between mirror matter and ordinary matter. Quantum effects could permit an ordinary photon to convert itself into a mirror photon on very rare occasions. If so, mirror matter would appear to have a small electrical charge.

26.   Some physicists insist that if the masses aren't identical, you shouldn't call it "mirror matter" but rather "shadow matter." The use of mirror matter seems to have become common in either case, however.

27.   Mohapatra, R. N., and V. Teplitz. 1996. Structures in the mirror universe. xxx.lanl.gov/abs/astro-ph/9603049, March 12.

28.   Astronomers have searched for MACHOs by training telescopes on the Magellanic Clouds, small satellite galaxies to the Milky Way. A Magellanic star brightens for a while when a MACHO passes in front of it (because the MACHO's gravity distorts the starlight). MACHO hunters have recorded about 20 cases of such Magellanic star brightenings. If the population of MACHOs in the Magellanic direction is typical, then they cannot account for all the dark matter estimated to lurk in the Milky Way's halo. By some estimates, MACHOs could make up half the invisible halo mass, but maybe a lot less, and certainly not all of it.

29.   Mohapatra, R. N., and V. Teplitz. 1999. Mirror matter MACHOs. xxx.lanl.gov/abs/astro-ph/9902085, February 4.

30.   M. Zapatero Osorio, et al. 2000. Discovery of young, isolated planetary mass objects in the σ Orionis star cluster. *Science* 290:103.

31.   Foot, R., et al. 2000. Do "isolated" planetary mass objects orbit mirror stars? xxx.lanl.gov/abs/astro-ph/0010502, October 25.

SUPER MATTER

1.   Hill, C., and L. Lederman. 2000. Teaching symmetry in the introductory physics curriculum. xxx.lanl.gov/abs/physics/0001061, version 2, February 7, pp. 1-2. See also www.emmynoether.com.

2.   Neal Lane, conversation with the author at Fermilab, June 14, 1999.

3.   McGrayne, S. B. 2001. *Nobel Prize Women in Science*, 2nd ed. Washington, D.C.: Joseph Henry Press, p. 72.

4.   Actually, the issue of conservation of energy in general relativity is more complicated than this; in different situations the very notions of energy and conservation are not easily defined.

5.   Technically, Noether showed that a conservation law is linked to a continuous symmetry. A sphere possesses continuous symmetry with respect to rotation, because it stays the same no matter how small a turn you give it. A snowflake has discrete symmetry, because you must turn it in increments of 60° to make it look the same. For more on this, see Hill and Lederman, pp. 6 ff.

6.   Another way of explaining it was suggested to me by Rabindra Mohapatra. If you rotate a triangle, all the points are changed at the same time, so the symmetry is "global." A gauge symmetry, on the other hand, allows changing a system at one point independently of other points. In a moving system where all points are connected, information about the change at one point must then be communicated to the other points; that communication is accomplished by the transmission of a force.

7.   Steven Weinberg, interview with the author in Austin, TX, November 21, 1997.

8.   Wilczek, F. 2001. Future summary. *International Journal of Modern Physics A* 16:1653-1678. Available at xxx.lanl.gov/abs/hep-ph/0101187.

9.   Edward Witten, interview with the author in Princeton, N.J., April 6, 1995.

10.   Ramond showed how fermions could be incorporated into string theory, paving the way for work showing the connection between string theory and supersymmetry.

11.   See Kane, G., and M. Shifman. 2000. Foreword. P. ix in *The Supersymmetric World: The Beginnings of the Theory*. Singapore: World Scientific. Also available at xxx.lanl.gov/abs/hep-ph/0102298.

12.   Savas Dimopoulos, conversation with the author in Houston, TX, November 1, 2000.

13.   Rita Bernabei, lecture in Austin, TX, December 11, 2000.

14.   Blas Cabrera, lecture in Austin, TX, December 11, 2000.

DARK MATTER

1. van den Bergh, S. 1999. The early history of dark matter. xxx.lanl.gov/abs/astro-ph/9904251, April 19, pp. 2-3.

2. Schucking, E. 2001. A personal memoir of 1958. *Physics Today* 54 (February):47.

3. Rosenfeld, L. 1967. Niels Bohr in the thirties. P. 127 in *Niels Bohr*, S. Rozental, ed. New York: John Wiley & Sons.

4. Pais, A. 2000. *The Genius of Science*. New York: Oxford University Press, p. 244.

5. Solomey, N. 1997. *The Elusive Neutrino*. New York: Scientific American Library, p. 16.

6. Reines, F. 1996. The neutrino: from poltergeist to particle. *Reviews of Modern Physics* 68(April):318.

7. Reines, p. 318.

8. The origin of the Q-ball idea seems to be a paper from 1985 by the Harvard physicist Sidney Coleman. The Q refers to a standard symbol that physicists use to denote a conserved quantity, or "charge." In this case the charge is a special symmetry property related to the particle number of the balls.

9. Kusenko, A. 1997. Q-balls in the MSSM. xxx.lanl.gov/abs/hep-ph/9707306, July 10.

10. Kusenko, A., and M. Shaposhnikov. 1997. Supersymmetric Q-balls as dark matter. xxx.lanl.gov/abs/hep-ph/9709492, version 3, October 30.

11. Dvali, G., et al. 1997. New physics in a nutshell, or Q-ball as a power plant. xxx.lanl.gov/abs/hep-ph/9707423, version 2, October 30.

12. Kolb, E., et al. 1998. WIMPZILLAS! xxx.lanl.gov/abs/hep-ph/9810361, October 14.

13. Rocky Kolb, interview with the author at Fermilab, May 8, 2001.

14. They are not necessarily colliding in the literal sense. After all, it is hard to say what it would mean for subatomic particles to collide, since they are not themselves like billiard balls but are rather fuzzy and wavy. By *collision* physicists mean that two particle come close enough to each other to cause some effect, such as a change in direction, which is pretty much what happens with real collisions, too.

THE BEST OF ALL POSSIBLE BUBBLES

1. Andrei Linde, interview by the author at The Woodlands, TX, January 7, 1991.

2. Rees, M. 1997. *Before the Beginning*. Reading, Mass.: Perseus Books, p. 3.

3. Tropp, E. A., et al. 1993. *Alexander A. Friedmann: The Man who Made the Universe Expand*. Cambridge: Cambridge University Press, p. 255.

4. Tropp et al, p. 73.

5. K. C. Cole and Rosie Mestel of the *Los Angeles Times*.

6. Tropp et al, p. 37.

7. The Dutch astronomer Willem de Sitter is sometimes credited with forecasting the expansion of the universe in a paper in 1917. While expansion may be implicit in de Sitter's work, he did not discuss it explicitly, as Friedmann did.

8. Friedmann, A. 1922. On the curvature of space. *Zeitschrift für Physik* 10:377-386. P. 49 in *Cosmological Constants*, J. Bernstein and G. Feinberg, eds. New York: Columbia University Press, 1986.

9. Friedmann, in Bernstein and Feinberg, p. 58.

10. Kragh, H. 1996. *Cosmology and Controversy*. Princeton: Princeton University Press, p. 27.

11. Hubble, E. 1929. A relation between distance and radial velocity among extragalactic nebulae. *Proceedings of the National Academy of Sciences*. 15(March 15):168-173. P. 81 in Bernstein and Feinberg.

12. At least it has always been interpreted as a slur. Hoyle apparently told a journalist years later that he merely meant to make the idea sound dramatic.

13. Gott, J. R. 2001. *Time Travel in Einstein's Universe*. Boston: Houghton Mifflin, p. 160.

14. Even before Guth's original version, a similar idea had been proposed in the Soviet Union by Alexei Starobinsky. But it was Guth's version that started the inflation bandwagon.

15. Alan Guth, lecture in Washington, D.C., April 14, 1999.

16. Rees, M. 2001. Concluding perspective. xxx.lanl.gov/abs/astro-ph/0101268, January 16, p. 6.

17. Rees, Concluding perspective, p. 8.

THE ESSENCE OF QUINTESSENCE

1. Lawrence Krauss, interview by the author in Washington, D.C., April 28, 2001.

2. Its official name is the Einstein field equation. Various textbooks present this equation in a wide range of different forms. This form is from one of Einstein's early papers: Einstein, A. 1915. On the general theory of relativity (addendum). *Sitzungsberichte der Königlich Preussischen Akademie der Wissenschaften zu Berlin*. P. 109 in *The Collected Papers*, vol. 6. Princeton: Princeton University Press, 1997.

3. Wheeler, J. 1990. *A Journey into Gravity and Spacetime*. New York: Scientific American Library, pp. 11-12.

4. The radiation density drops off more rapidly because it depends on the fourth power of the universe's radius. The matter density depends only on the cube of the radius—in other words, it's proportional to the volume. So radiation density drops faster than matter density and at some point will fall below the matter density.

5. "Initial conditions" are a necessary part of applying the laws of physics. The laws simply state what will happen, given such-and-such a situation—the position and velocities of particles, any forces in the neighborhood, etc. Those are the initial conditions.

6. Einstein, A. 1917. Cosmological considerations on the general theory of relativity. *Sitzungsberichte der Preussischen Akademie der Wissenschaften*. P. 26 in *Cosmological Constants*, J. Bernstein and G. Feinberg, eds. New York: Columbia University Press, 1986.

7. Gamow, G. 1970. *My World Line*. New York: Viking Press, p. 44.

8. In 1923, Hermann Weyl showed how de Sitter's approach to Einstein's theory could describe an expanding universe. "If there is no quasi-static world, then away with the cosmological term," Einstein replied to Weyl in a postcard. But Einstein continued to reject the idea of an expanding universe as physically real, as he told Lemaître at a meeting in 1927. Only after Hubble's analysis did Einstein explicitly disavow the cosmological constant, in an obscure journal in 1931. See Straumann, N. 2000. On the mystery of the cosmic vacuum energy density. xxx.lanl.gov/abs/astro-ph/0009386, September 25, pp. 3-4.

9. Einstein, A. 1956. *The Meaning of Relativity*, 5th ed. Princeton: Princeton University Press, p. 127.

10. Josh Frieman, discussion with reporters at Fermilab, May 1, 1998.

11. Siegfried, T. 1992. Einstein buried his "mistake," but it's still haunting scientists. *Dallas Morning News*, January 20, p. 7D.

12. Discussion with reporters at Fermilab, May 1, 1998.

13. Michael Turner, lecture at Fermilab, May 1, 1998.

14. In case you missed it, these were figure skater Nancy Kerrigan's comments after she was bashed in the knee shortly before the 1994 Winter Olympics.

15. Other precursor papers can be traced back to the early 1980s. Frieman, by the way, does not like the name quintessence, arguing tongue-in-cheek that *pentessence* would make more sense, because Aristotle was Greek, not Roman.

16. Robert Caldwell, talk at Fermilab, May 1, 1998.

17. Michael Turner, interview by the author at Fermilab, May 26, 1999.

18. Gu, J.-A., and W.-Y. P. Hwang. 2001. The fate of the accelerating universe. xxx.lanl.gov/abs/astro-ph/0106387, June 21.

19. Krauss, L. 2001. *Atom*. Boston: Little, Brown, p. 275.

20. Lawrence Krauss, interview by the author in Washington, D.C., April 28, 2001.

SUPERSTRINGS

1. Newman, J. 1955. James Clerk Maxwell. *Scientific American* 192(June). P. 156 in *Lives in Science*. New York: Simon & Schuster.

2. Siegel, D. 1981. Thomson, Maxwell, and the universal ether in Victorian physics. P. 249 in *Conceptions of Ether*, G. N. Cantor and M. J. S. Hodge, eds., Cambridge: Cambridge University Press.

3. Siegel, p. 254.

4. Maxwell, J. C. 1864. A dynamical theory of the electromagnetic field. *Philosophical Transactions of the Royal Society of London* 155. P. 857 in *The World of Physics*, vol. 1, by J. H. Weaver. New York: Simon & Schuster, 1987.

5. Holton, G. 1971-1972. On trying to understand scientific genius. *American Scholar* 41(Winter):102.

6. Stachel, J. 1998. *Einstein's Miraculous Year*. Princeton: Princeton University Press, p. 15.

7. Schwarz, J. H. 2000. String theory: the early years. xxx.lanl.gov/abs/hep-th/0007118 version 3, July 26, p. 3.

8. Murray Gell-Mann, interview by the author in Santa Fe, N.M., September 16, 1997.

9. Schwarz, J. H. 2000. Reminiscences of collaborations with Joël Scherk. xxx.lanl.gov/abs/hep-th/0007117, July 14, p. 3.

10. Schwarz, Reminiscences, p. 4.

11 Both correctly point out that such a theory would not explain all the events that depend on historical contingency or any of a number of complicated things. But, while acknowledging their legitimate objections, I use the phrase occasionally as a convenient shorthand.

12. Sullivan, W. 1985. Is absolutely everything made of string? *New York Times*, May 7, p. 1C.

13. Siegfried, T. 1985. Superstring: theory ties forces together in major physics breakthrough, *Dallas Morning News*, April 22, p. 7D.

14. To write that number out, you'd put 30 zeroes to the right of the decimal point, then the 1.

15. Murray Gell-Mann, interview in Santa Fe, N.M., September 16, 1997.

16. Mach, E. 1960. *The Science of Mechanics*. LaSalle, Ill.: Open Court Publishing, pp. 588-589.

17. Mach, E. 1960. *Space and Geometry*. LaSalle, Ill.: Open Court Publishing, p. 138.

STRETCHING YOUR BRANE

1. Rocky Kolb, interview with the author at Fermilab, June 16, 1999.

2. Joe Lykken, conversation with the author at Fermilab, June 15, 1999.

3. Israel, W. 1987. Dark stars: the evolution of an idea. P. 201 in *300 Years of Gravitation*, S. W. Hawking and W. Israel, eds. Cambridge: Cambridge University Press. Michell's paper was communicated by his friend Cavendish to the Royal Society on November 27, 1783.

4. Israel, p. 203.

5. Thorne, K. S. 1994. *Black Holes and Time Warps*. New York: W. W. Norton, p. 124.

6. Bernstein, J. 1996. The reluctant father of black holes. *Scientific American* 274(June):83.

7. Schwarzschild, K. 1916. On the gravitational field of a sphere of incompressible fluid according to Einstein's theory. *Sitzungsberichte der Königlich Preussischen Akademie der Wissenschaften zu Berlin* (1916), translated by S. Antoci. xxx.lanl.gov/abs/physics/9912033, December 16, 1999, p. 9. To be precise, Schwarzschild wrote, "For an observer measuring from the outside . . . a sphere of given gravitational mass $\alpha/2K^2$ can not have a radius measured from the outside smaller than $P_0 = \alpha$." For a sphere of incompressible fluid, the limit is 9/8 times α.

8. Bernstein, p. 84.

9. Oppenheimer, J. R., and H. Snyder. 1939. On continued gravitational contraction. *Physical Review* 56(September 1):457.

10. Oppenheimer and Snyder, p. 456.

11. Oppenheimer and Snyder, p. 459.

12. The Oppenheimer-Snyder paper was in the same issue as the famous Bohr-Wheeler paper describing the basic physics of nuclear fission.

13. Thorne, pp. 210-211.

14. Siegfried, T. 1998. Black hole was catchy for Wheeler, *Dallas Morning News*, October 19, p. 4F.

15. Newcomb, S. 1894. Modern mathematical thought. *Nature* 49:325-329. P. 386 in *Time Machines*, by P. Nahin. 2nd ed. New York: Springer-Verlag.

16. Isaksson, E. Gunnar Nordström (1881-1923): on gravitation and relativity. www.helsinki.fi/~eisaksso/nordstrom/nordstrom.html.

17. Kaluza, T. 1921. On the unification problem of physics. *Sitzungsberichte der Königlich Preussischen Akademie der Wissenschaften zu Berlin*. P. 53 in *The Dawning of Gauge Theory*, L. O'Raifeartaigh, ed. Princeton: Princeton University Press.

18. Klein, O. 1926. Quantum theory and five-dimensional relativity. *Zeitschrift für Physik* 37. P. 68 in O'Raifeartaigh.

19. Instead of obeying the inverse-square law, for instance, the strength of gravity would diminish as the cube of the distance between two bodies, assuming one additional dimension.

20. Andy Strominger, telephone interview by the author, 1995.

21. Later, Duff moved to the University of Michigan.

22. Siegfried, T. 1990. Superstrings snap back, *Dallas Morning News*, March 19, p. 6D.

23. I encountered the triangle-cone example in Durham, I. T. 2000. A historical perspective on the topology and physics of hyperspace. xxx.lanl.gov/abs/physics/0011042, November 18.

24. Savas Dimopoulos, interview by the author in Palo Alto, Calif., February 20, 2001.

25. In that view, the boundary is 10-dimensional, but maybe only three of the space dimensions are big, so that our universe appears to us to be a three-brane.

26. These parallel worlds are not the same thing as the multiverse, the multiple bubbles of spacetime inflating out of a common vacuum. The multiple bubbles we met before would all be just parts of our own familiar three-dimensional space—too far away to communicate with, but part of our same fabric. They would be very, very distant—too far away for light to ever travel from there to here. In other words, there's no need to worry about what's going on in them. But the parallel brane worlds could literally be less than a silly millimeter away.

27. Joe Lykken, telephone interview by the author, July 1, 1999.

28. Lisa Randall, interview by the author in Ann Arbor, Mich., July 13, 2000.

29. Joe Lykken, interview by the author in Lake Tahoe, Calif., December 11, 1999.

30. Rocky Kolb, interview by the author at Fermilab, June 16, 1999.

31. Lisa Randall, talk in San Francisco at the annual meeting of the American Association for the Advancement of Science, February 16, 2001.

32. Joe Lykken, interview by the author in Lake Tahoe, Calif., December 11, 1999.

GHOSTS

1. One modern commentator's assessment: "Kant's belief that Euclidean geometry was true, because our intuitions tell us so, seems to me to be either unintelligible, or wrong." P. 85 in Gray, J. 1989. *Ideas of Space*. Oxford: Clarendon Press.

2. Kline, M. 1985. *Mathematics and the Search for Knowledge*. New York: Oxford University Press, p. 152.

3. Kline, p. 152.

4. Kline, p. 152.

5. Bell, E. T. 1937. *Men of Mathematics*. New York: Simon & Schuster, p. 297.

6. Weber, A. S., ed. 2000. *19th Century Science: An Anthology*. Peterborough, Canada: Broadview Press, p. 138.

7. Bolyai's comment came in a letter to his father in 1823, quoted (with a slightly different translation) on p. 107 of *Ideas of Space*, by J. Gray. Oxford: Clarendon Press, 1989.

8. Bell, p. 490.

9. Riemann, B. 1959. On the hypotheses which lie at the foundations of geometry. P. 411 in *A Source Book in Mathematics*, by D. E. Smith. New York: Dover Publications.

10. Smith, p. 424.

11. Smith, p. 425.

12. Einstein's main problem in formulating general relativity was to find a mathematical way of expressing "general covariance"—the equivalence of all accelerating systems regardless of the coordinate system you used to keep track of their motion. But Einstein could not reconcile general covariance with Euclidean geometry. Grossmann showed Einstein that Riemannian geometry could describe general covariance consistently.

13. Einstein, A. 1915. On the general theory of relativity (addendum). *Sitzungsberichte der Königlich Preussischen Akademie der Wissenschaften zu Berlin*. P. 108 in *The Collected Papers*, vol. 6. Princeton: Princeton University Press, 1997.

14. Keep in mind, Einstein's general relativity describes the curvature of space and time combined, as spacetime. But talking only about the curvature of space is a shorthand approach that usually does no damage.

15. Luminet, J.-P., and B. Roukema. 1999. Topology of the universe: theory and observation. xxx.lanl.gov/abs/astro-ph/9901364, January 26, p. 2.

16. Luminet and Roukema, pp. 2-4.

17. News conference, American Physical Society meeting, Columbus, Ohio, April 17, 1998.

18. Levin, J. 2001. Topology and the cosmic microwave background. xxx.lanl.gov/abs/gr-qc/0108043, August 16, p. 3.

19. David Spergel, interview by the author in Chapel Hill, N.C., April 11, 2001.

20. Luminet, J.-P., et al. 1999. Is space finite? *Scientific American* 280(April):92.

21. Barrow, J., and J. Levin. 1999. Chaos and order in a finite universe. xxx.lanl.gov/abs/astro-ph/9907288, July 21, p. 2.

22.   Levin, J., and I. Heard. 1999. Topological pattern formation. xxx.lanl.gov/abs/
       astro-ph/9907166, July 13, p. 1.

THE TWO-TIMING UNIVERSE

1.   Clark, R.W. 1971. *Einstein: The Life and Times*. New York: World Publishing, p. 10.
2.   Einstein, A. 1951. Autobiographical notes. P. 5 in *Albert Einstein: Philosopher-Scientist*, P. Schilpp, ed., vol. 1. New York: Harper and Row.
3.   Einstein, p. 16.
4.   Einstein, p. 53.
5.   If you are moving in a straight line at constant speed, you occupy a legitimate frame of reference, or "inertial frame," for making observations and measurements. Your conclusions about the laws of physics should be the same as those of anyone else in any other inertial frame. Therefore there must be some symmetry group—a group of operations—that can reorient your inertial frame to make it identical to any other inertial frame. The mathematical operations describing that symmetry are known as the Lorentz group.
6.   If an object somehow begins life at faster-than-light speeds, special relativity's rules would not be broken. Such particles, known as tachyons, would be "legal," but there is no solid evidence that they actually exist. There is another possible loophole to the speed-of-light limit, known as the Scharnhorst effect, in which light can go slightly faster than its usual speed in a vacuum. The Scharnhorst effect achieves this trick by putting two metal plates close enough together to restrict the wavelengths of photons that can pop into existence out of the vacuum. This effect "clears out" some of the quantum clutter in the vacuum, enabling light to zip through more rapidly. The extra speed is far too small to measure, though. If you started a race between a Scharnhorst photon and an ordinary photon at the time of the universe's birth, by now the faster photon would be ahead by less than the width of an atom.
7.   Einstein, A. 1905. On the electrodynamics of moving bodies. *Annalen der Physik* 17. P. 139 in *Einstein's Miraculous Year*, by J. Stachel. Princeton: Princeton University Press, 1998.
8.   Einstein, A. 1911. The Theory of Relativity, lecture in Zurich, January 16, 1911. *Naturforschende Gesellschaft in Zürich. Vierteljahrsschrift* 56. Pp. 348-349 in *The Collected Papers*, vol. 3. Translated by A. Beck. Princeton: Princeton University Press, 1993.
9.   Several books offer good in-depth explanations of the twin paradox. You might try Davies, P. 1995. *About Time*. New York: Simon & Schuster; Greene, B. 1999. *The Elegant Universe*. New York: W. W. Norton; or pp. 462 ff in Nahin, P. 1999. *Time Machines*. 2nd ed. New York: Springer-Verlag.
10.   Minkowski, H. 1908. Space and Time, address delivered at the 80th Assembly of German Natural Scientists and Physicians, at Cologne, September 21, 1908. P.75 in *The Principle of Relativity*, by A. Einstein, et al., translated by W. Perrett and G. B. Jeffery. New York: Dover Publications, 1952.

11.   Nahin, P. 1999. *Time Machines*. 2nd ed. New York: Springer-Verlag, pp. 140 ff.
12.   Cumrun Vafa, e-mail correspondence with the author, October 23, 1996.
13.   Andy Strominger, telephone interview with the author, October 22, 1996.
14.   Cumrun Vafa, e-mail correspondence with the author, October 23, 1996.
15.   Tegmark, M. 1997. On the dimensionality of spacetime. xxx.lanl.gov/abs/gr-qc/9702052, version 2, April 4, p. 2.
16.   Tegmark, p. 3.
17.   Bars, I., and C. Kounnas. 1997. Theories with two times. xxx.lanl.gov/abs/hep-th/9703060, March 7.
18.   Hull, C. M. 1999. Duality and strings, space, and time. xxx.lanl.gov/abs/hep-th/9911080, November 11, pp. 3-4.
19.   Hull, p. 12.
20.   Hull, p. 14.
21.   Hull, p. 14.

EPILOGUE

1.   Lippmann, W. 1936. *Public Opinion*. New York: Macmillan, p. 25.
2.   Lippmann, p. 29.
3.   Wigner, E. 1991. The unreasonable effectiveness of mathematics in the natural sciences. *Communications in Pure and Applied Mathematics*. P. 527 in *The World Treasury of Physics, Astronomy, and Mathematics*, T. Ferris, ed. Boston: Little, Brown.
4.   Lakoff, G., and R. Núñez. 2000. *Where Mathematics Comes From*. New York: Basic Books, p. 9.
5.   George Lakoff, talk in San Francisco at the annual meeting of the American Association for the Advancement of Science, February 17, 2001.
6.   Hertz, H. 1945. On the relations between light and electricity. P. 459 in *The Autobiography of Science*, F. R. Moulton and J. J. Schifferes, eds. Garden City, N.Y.: Doubleday, Doran.
7.   Murray Gell-Mann, dinner talk in Pasadena, Calif., January 14, 2000.
8.   Rocky Kolb, interview by the author in Chapel Hill, N.C., April 13, 2001.
9.   Hertz, p. 460.
10.   Holton, G. 1971-1972. On trying to understand scientific genius. *American Scholar* 41(Winter):102.
11.   Cumrun Vafa, e-mail correspondence with the author, October 23, 1996.
12.   For an elaboration on this view with regard to Spinoza, see Zimmermann, R. E. 2000. Loops and knots as topoi of substance. Spinoza revisited. xxx.lanl.gov/abs/gr-qc/0004077, version 2, May 23.
13.   Poincaré, H. 1952. *Science and Hypothesis*. New York: Dover Publications, p. 158.
14.   Poincaré, p. 159.
15.   Poincaré, p. 161.
16.   Poincaré, H. 1963. *Mathematics and Science: Last Essays*, translated by J. W. Bolduc. New York: Dover Publications, p. 14.
17.   Planck, M. 1949. *Scientific Autobiography and Other Papers*. New York: Philosophical Library, p. 58.

18. Poincaré, *Mathematics and Science*, p. 13.
19. Poincaré, *Science and Hypothesis*, p. 161.
20. As I was reading the proofs for this book, I was also reading the proofs of a new book by Stephen Wolfram, the physicist-turned-entrepreneur famous for developing the computer program Mathematica. Wolfram's book, called *A New Kind of Science*, offers some interesting insights into the relationship of math to reality. By studying computer programs called cellular automata, Wolfram demonstrates that very simple rules can produce structures of great complexity. In fact, he shows, programs that exhibit behavior beyond some minimum threshold of complexity, while still fairly simple, can emulate any other computing system of whatever complexity. He therefore deduces a "principle of computational equivalence," declaring that any programs or natural processes exceeding that threshold are ultimately equivalent in their computational sophistication.

   In Wolfram's view all natural processes can be considered to be, in essence, computations. And obviously mathematics can also be regarded as computation as well. Wolfram therefore concludes that there is an intrinsic equivalence between nature and mathematics, as all computation that is not trivially simple possesses equivalent sophistication. The power of math to represent reality is therefore merely a reflection of the intrinsic equivalence of both math and reality as equally powerful forms of computation.

21. Lippmann, p. 16.
22. Lippmann, p. 29.

FURTHER READING

INTRODUCTION

Abbott, Edwin A. *Flatland*. New York: Dover, 1952.

Abbott, Edwin A. *The Annotated Flatland*. With introduction and notes by Ian Stewart. Cambridge, Mass.: Perseus, 2002.

Wigner, Eugene. "The Unreasonable Effectiveness of Mathematics in the Natural Sciences," *Communications in Pure and Applied Mathematics*, reprinted in Ferris, T., ed., *The World Treasury of Physics, Astronomy, and Mathematics*. Boston: Little, Brown and Company, 1991, pp. 526-540.

Lakoff, George and Núñez, Rafael. *Where Mathematics Comes From*. New York: Basic Books, 2000.

STRANGE MATTER

Crawford, Henry J. and Greiner, Carsten H. "The Search for Strange Matter." *Scientific American* 270 (January 1994), 72-77.

Pais, Abraham. *Inward Bound*. New York: Oxford University Press, 1986.

MIRROR MATTER

Dirac, P. A. M. *Directions in Physics*. New York: John Wiley & Sons, 1978.

Pais, Abraham. *Inward Bound*. New York: Oxford University Press, 1986.

Pais, Abraham. *The Genius of Science*. New York: Oxford University Press, 2000.

SUPER MATTER

Byers, Nina. "E. Noether's Discovery of the Deep Connection between Symmetries and Conservation Laws." xxx.lanl.gov/abs/physics/9807044, version 2, September 23, 1998.

Kane, Gordon. *Supersymmetry*. Cambridge, Mass.: Perseus, 2000.

Weyl, Hermann. *Symmetry*. Princeton: Princeton University Press, 1952.

DARK MATTER

Solomey, Nickolas. *The Elusive Neutrino*. New York: Scientific American Library, 1997.

THE BEST OF ALL POSSIBLE BUBBLES

Guth, Alan. *The Inflationary Universe*. Reading, Mass.: Helix Books, 1997.

Rees, Martin. *Before the Beginning*. Reading, Mass.: Perseus, 1997.

Tropp, E. A., Frenkel, V. Ya., and Chernin, A. D. *Alexander A. Friedmann: The Man who Made the Universe Expand*. Cambridge: Cambridge University Press, 1993.

THE ESSENCE OF QUINTESSENCE

Caldwell, Robert and Steinhardt, Paul. "Quintessence." *Physics World* 13 (November 2000), 31-37.

Krauss, Lawrence. *Quintessence*. New York: Basic Books, 2000

SUPERSTRINGS

Davies, P. C. W. and Brown, J. R., eds. *Superstrings*. Cambridge: Cambridge University Press, 1988.

Greene, Brian. *The Elegant Universe*. New York: Norton, 1999.

Newman, James. "James Clerk Maxwell." *Scientific American* 192(June 1955), 58-71. Reprinted in *Lives in Science*. New York: Simon and Schuster, 1957, pp. 155-180.

STRETCHING YOUR BRANE

Abel, Steven and March-Russell, John. "The Search for Extra Dimensions." *Physics World* 13 (November 2000), 39-44.

Bernstein, Jeremy. "The Reluctant Father of Black Holes." *Scientific American* 274(June 1996), 80-85.

Duff, Michael. "The Theory Formerly Known as Strings." *Scientific American* 278(February 1998), 64-69.

Greene, Brian. *The Elegant Universe*. New York: W. W. Norton, 1999.

Thorne, Kip S. *Black Holes and Time Warps*. New York: W. W. Norton, 1994.

GHOSTS

Gray, Jeremy. *Ideas of Space*. Oxford: Clarendon Press, 1989.

Kline, Morris. *Mathematics and the Search for Knowledge*. New York: Oxford University Press, 1985.

Luminet, Jean-Pierre, Starkman, Glenn, and Weeks, Jeffrey. "Is Space Finite?" *Scientific American* 280(April 1999), 90-97.
Monastyrsky, Michael. *Riemann, Topology, and Physics.* Second Edition. Translated by Roger Cooke, James King, and Victoria King. Boston: Birkhauser, 1999.

THE TWO-TIMING UNIVERSE

Hull, C. M. "Duality and Strings, Space and Time." xxx.lanl.gov/abs/hep-th/9911080, November 11, 1999.
Nahin, Paul. *Time Machines.* Second Edition. New York: Springer-Verlag, 1999.
Stachel, John. *Einstein's Miraculous Year.* Princeton: Princeton University Press, 1998.

EPILOGUE

Poincaré, Henri. *Mathematics and Science: Last Essays.* Translated by J. W. Bolduc. New York: Dover, 1963.
Poincaré, Henri. *Science and Hypothesis.* New York: Dover, 1952.

INDEX